DevOps Simplified: Zero-Maintenance Strategies for AWS EKS

Efficient Deployment
and Management Strategies
for AWS EKS Environments
with Terraform

Dmytro Kozhevin

Apress®

DevOps Simplified: Zero-Maintenance Strategies for AWS EKS: Efficient Deployment and Management Strategies for AWS EKS Environments with Terraform

Dmytro Kozhevin
Kyiv, Ukraine

ISBN-13 (pbk): 979-8-8688-1057-2   ISBN-13 (electronic): 979-8-8688-1058-9
https://doi.org/10.1007/979-8-8688-1058-9

Copyright © 2024 by Dmytro Kozhevin

This work is subject to copyright. All rights are reserved by the Publisher, whether the whole or part of the material is concerned, specifically the rights of translation, reprinting, reuse of illustrations, recitation, broadcasting, reproduction on microfilms or in any other physical way, and transmission or information storage and retrieval, electronic adaptation, computer software, or by similar or dissimilar methodology now known or hereafter developed.

Trademarked names, logos, and images may appear in this book. Rather than use a trademark symbol with every occurrence of a trademarked name, logo, or image we use the names, logos, and images only in an editorial fashion and to the benefit of the trademark owner, with no intention of infringement of the trademark.

The use in this publication of trade names, trademarks, service marks, and similar terms, even if they are not identified as such, is not to be taken as an expression of opinion as to whether or not they are subject to proprietary rights.

While the advice and information in this book are believed to be true and accurate at the date of publication, neither the authors nor the editors nor the publisher can accept any legal responsibility for any errors or omissions that may be made. The publisher makes no warranty, express or implied, with respect to the material contained herein.

>   Managing Director, Apress Media LLC: Welmoed Spahr
>   Acquisitions Editor: Celestin Suresh John
>   Development Editor: Gryffin Winkler
>   Coordinating Editor: Gryffin Winkler

Cover designed by eStudioCalamar

Cover image by Lukasz Lada @ Unsplash.com

Distributed to the book trade worldwide by Apress Media, LLC, 1 New York Plaza, New York, NY 10004, U.S.A. Phone 1-800-SPRINGER, fax (201) 348-4505, e-mail orders-ny@springer-sbm.com, or visit www.springeronline.com. Apress Media, LLC is a California LLC and the sole member (owner) is Springer Science + Business Media Finance Inc (SSBM Finance Inc). SSBM Finance Inc is a **Delaware** corporation.

For information on translations, please e-mail booktranslations@springernature.com; for reprint, paperback, or audio rights, please e-mail bookpermissions@springernature.com.

Apress titles may be purchased in bulk for academic, corporate, or promotional use. eBook versions and licenses are also available for most titles. For more information, reference our Print and eBook Bulk Sales web page at http://www.apress.com/bulk-sales.

Any source code or other supplementary material referenced by the author in this book is available to readers on GitHub (https://github.com/Apress). For more detailed information, please visit https://www.apress.com/gp/services/source-code.

If disposing of this product, please recycle the paper

Table of Contents

About the Author xi
About the Technical Reviewer xiii
Introduction xv

Chapter 1: Setting Our Sights 1
Thinking in Layers for Accelerated Development.. 1
    Foundation Layer 1
    Kubernetes Core Resources Layer 1
    CI/CD Layer 2
    Application Layer 2
Directory Structure.. 2
A Few Words About Uniqueness... 3
Terraform 101 3
    The Early Days of Infrastructure 3
    The Declarative Revolution 4
    Declaring a Resource 4
    State Management 4
    Providers 5
    Fetching Data 5
    Local Variables 6
    Input Variables 6
    Outputs 7
    Organizing Your Working Directory 7

TABLE OF CONTENTS

Chapter 2: Starting from Scratch 9
Beginning of the Journey: VPC, DNS, and ECR 9
The Current Phase of the Organization 9
DevOps Goals 9
Minimal Required Work via Web Console 10
Setting Up CLI 10
Creating State Bucket via CLI 11
First Steps in Our Infra Git Repo 11
Variables in variables.tf 11
Main.tf 12
Building the VPC (Virtual Private Cloud) 13
Dealing with DNS (Domain Name System) 15
ECR (Elastic Container Registry) 17

Chapter 3: Building the Infrastructure Base 19
Our First IAM Role: AWS Identity and Access Management 101 19
How I Expected It to Be 19
AWS IAM Is Not RBAC 19
The Role of a Role 20
We Need an IAM Policy for Our Valet 21
Karpenter Policy 21
Dramatically Reducing Compute Bills 26

Chapter 4: Sculpting Kubernetes: The Initial Steps 29
Introducing Kubernetes 29
How Would You Do That? 29
Kubernetes Database: Etcd 29
A Control Hub 31

It's a Lot of Servers..33

One to Control Everything Kubernetes Controller Manager.........................35

Where Are the Real Workers? Master vs. Worker Nodes 39

Chapter 5: Bridging Kubernetes and AWS Resources........................43

Elastic Kubernetes at Your Service 43

    The Standard Kubernetes Cluster..43

    Setting Up a Kubernetes Cluster in AWS 43

    Our Choice: EKS...44

    Our Choice: EKS by Terraform..44

    Our Choice: EKS by Terraform EKS Module...45

    Get to Work 45

    Summary of Level 0...48

    It's Time to Make It Alive..48

    Custom Nameservers for Domain..48

Chapter 6: The Structure of Application 49

Common Kubernetes Resources Part 1 49

    Introduction Level 1 49

    Variables..49

    Main.tf with State and Providers..50

    Explanation of the Code in main.tf 52

    The Power of EKS Add-ons 53

    Making Our Lives Easier with Helm...55

    How Does Helm Work? 55

    Key Components of Helm 55

    Why Use Helm? 56

    Why We Use Helm from Terraform Instead of the Helm CLI..........................56

    In Summary 57

TABLE OF CONTENTS

Chapter 7: Be a Good Kubernetes Citizen 59

Common Kubernetes Resources Part 2 59

    What Is a Storage Class? 59

    Why Do We Need a Custom Storage Class for EBS? 60

    Terraform Code for Kubernetes Storage 60

    Why Not Use the Default gp2 Storage Class? 63

    Conclusion on Storage 64

    Karpenter Application 64

    Summary on Karpenter 68

    Why Use an Ingress Object? 68

    What Is an Ingress Controller? 68

    Why Use the ALB Ingress Controller? 69

    Let's Install ALB Ingress Controller 69

    External DNS 70

    It's Time to Apply 74

Chapter 8: The Intricacies of Helm 75

The Hard Parts 75

    Kubectl: Your Key to the Kubernetes Cluster 75

    Kubectl Configuration 76

    Basic Kubectl Commands 76

    Kubectl api-resources 77

    Introducing CRDs 77

    Terraform's Inconvenience with CRDs 77

    Creating the 02-karpenter Folder 78

    The Secret of aws-auth ConfigMap 78

Understanding the Content of aws-auth 79

TABLE OF CONTENTS

How to Read and Edit the aws-auth ConfigMap 80
    Time to Create Main.tf 80
Backend Configuration...80
Cluster Information and Kubernetes Provider...81
IAM Role for Karpenter Nodes...82
Attaching Policies to the IAM Role 84
Creating Kubernetes Resources for Karpenter 85
Creating EC2NodeClass Resource for Karpenter 88
Updating the aws-auth ConfigMap 89
    Put in Action 90

Chapter 9: Choosing the Right CI/CD Platform 91

Setting GitLab in Action..91
    Why GitLab?..91
    GitLab Components 92
    Use Helm to Install GitLab 92
    Let's Start 93
    Put main.tf 93
    variables.tf..94
    outputs.tf...95
    Terraform Template Files 95
    gitlab-values.tpl...96
    Understanding the GitLab Values File 97
    gitlab.tf 101
    Terraform Apply 104
    Possible Issues 105
    Conclusion 105

vii

TABLE OF CONTENTS

Chapter 10: Practical Example Part 1 107
Introduction 107
A Learning Environment 107
A New Folder 108
Content of variables.tf 108
Content of main.tf 109
Setting Up User Access 111
AWS User Creation 111
AWS Access Key 112
Every User Gets an IAM Policy Attached 112
RBAC: Role-Based Access Control for Students 113
Access to Kubernetes 119

Chapter 11: Practical Example Part 2 123
DNS Records 123
Docker Image 125
Build and Push Docker Image 125
Namespaces for WebIDE 126
Persistent Storage for Student Work 126
Service for WebIDE 127
Introducing Ingress 128
Creating Ingress for Every Student 129
Init Script 131
Random Password 132

TABLE OF CONTENTS

Chapter 12: Practical Example Part 3 133

Deploying the WebIDE with StatefulSet 133

Access Details Distribution 139

Providing List of Students 140

Further Improvements 140

Time to Apply 141

Conclusion 141

Index 143

About the Author

Dmytro Kozhevin is an accomplished DevOps engineer, educator, and robotics enthusiast with over 18 years of experience. He now channels his expertise into building a DevOps practice platform and advancing robotics applications. Dmytro's career has been marked by a relentless commitment to innovation in cloud-native solutions, with a focus on Kubernetes and scalable cloud infrastructures on AWS and GCP. Passionate about simplifying complex DevOps challenges, he delivers accessible, practical solutions to help teams excel in the rapidly evolving tech landscape.

About the Technical Reviewer

Mohammed Ilyas Ahmed is an industry professional with extensive expertise in security within the DevSecOps domain, where he diligently works to help organizations bolster their security practices. With a fervent dedication to enhancing security posture, Mohammed's insights and guidance are invaluable to those navigating the complex landscape of DevSecOps. In addition to his involvement in industry events, Mohammed is an active speaker and judge, lending his expertise to technical sessions at prestigious conferences. His commitment to advancing knowledge is evident through his research contributions at Harvard University, where he contributes to journal publications, enriching the academic discourse surrounding security practices, and, as a distinguished member of the Harvard Business Review Advisory Council, underscores his commitment to advancing knowledge and fostering collaboration between academia and industry. He is also the author of the book *Cloud-Native DevOps*.

Mohammed Ilyas Ahmed's influence extends even further as a member of the Global Advisory Board at VigiTrust Limited, based in Dublin, Ireland. This additional role highlights his international reach and his involvement in shaping global strategies for cybersecurity and data protection.

Mohammed's dedication to excellence is further highlighted by his numerous certifications, which serve as a testament to his proficiency and depth of knowledge in the security domain. However, beyond his professional pursuits, Mohammed is a multifaceted individual with a diverse range of interests, adding richness to his character and perspective.

Introduction

Welcome to *DevOps Simplified: Zero-Maintenance Strategies for AWS EKS*. In this guide, you'll uncover over ten years of hands-on experience with AWS, Terraform, Kubernetes, and Helm, distilled into practical insights and strategies from countless DevOps training sessions and real-world applications. This book is designed for

1. **CTOs and Team Leads:** Gain insights into setting up and maintaining a robust CI/CD pipeline in a rapidly scaling startup. This book guides you through building a resilient infrastructure that requires minimal maintenance—ideal for environments not yet ready for a full-time DevOps team.
2. **Aspiring DevOps Engineers:** Embark on a pragmatic journey through AWS EKS. Filled with practical advice, real-world scenarios, and strategies, this book is an invaluable resource for those beginning their path in DevOps.

Main Focus

The core of this book is to provide you with

- **Working Terraform Code:** Dive into executable Terraform code, designed for real-world applications.

INTRODUCTION

- **Practical Helm Examples:** Explore practical Helm chart examples that you can implement and modify as needed.
- **Line-by-Line Explanation:** Each code snippet is accompanied by detailed explanations, clarifying the rationale behind every decision and line of code.

Now, let's talk about the big picture—and by big picture, I mean building an infrastructure that makes you feel like you're playing with the best tools in the DevOps toolbox. Imagine you're at the helm of a startup that's just starting out—no real users yet, no production environment, but with dreams as big as the cloud you're building on. Your challenge? To create a cost-effective setup that not only keeps AWS bills delightfully low but also minimizes the need for a full-time DevOps engineer. You want to set it up, let it hum along smoothly for a year, and check back only when you want to—not because you have to.

This infrastructure isn't just about saving money; it's about having the freedom to experiment and innovate at lightning speed. Need a new environment for a demo or a quick experiment? No problem—spin it up and tear it down with ease, all without breaking a sweat or your budget.

This book isn't just another manual—it's a dense, no-nonsense playbook, honed over a year of writing and a decade of hands-on experience. It's designed to be your go-to resource, sitting right on your desk, ready to help you navigate the challenges of the next year and beyond. Inside, you'll find the distilled essence of what it took to build hundreds of Kubernetes clusters—knowledge that I've refined through countless trials, and now you get to leverage it directly.

INTRODUCTION

Here's the deal: even though we all use the same tools—Terraform, Kubernetes, Helm—no two infrastructures are ever the same. That's the beauty of this work; it's both art and science. This book doesn't just hand you cookie-cutter solutions—it empowers you to carve out your own path, to be unique, while standing on the shoulders of proven patterns that work.

Whether you're a startup scaling at breakneck speed or a DevOps pro sharpening your skills, this guide is packed with the tools, strategies, and insights you need to streamline your DevOps practices. You're not just getting a book—you're getting a decade's worth of hard-earned wisdom, all condensed into a comprehensive guide designed to help you build, innovate, and succeed.

CHAPTER 1

Setting Our Sights

Thinking in Layers for Accelerated Development

Imagine constructing your infrastructure like a multi-layered cake. This metaphor isn't just for culinary enthusiasts; it's a practical framework for organizing your repository's folder structure, selecting the appropriate tools, and enhancing security right from the start.

Foundation Layer

At the base, we establish the elements we seldom modify—those aspects of our infrastructure that are fundamental and enduring. This includes the virtual private cloud (VPC), IAM roles, EKS cluster, S3 buckets, Elastic Container Registry (ECR), and RDS. We use Terraform to define and manage these resources, anticipating changes only once or twice a year.

Kubernetes Core Resources Layer

Moving up, the next layer contains Kubernetes resources shared across all our products. Here, we find essential components like Cert Manager, Ingress Controller, and the Prometheus Operator, complemented by

Loki and Grafana for monitoring. These are elements we might adjust a few times a year. For this layer, we blend Terraform and Helm, balancing flexibility with stability.

CI/CD Layer

This layer houses a living GitLab cluster, vital for continuous integration and deployment.

Application Layer

The application layer, while highly dynamic, primarily concerns deploying and updating applications in a production-like environment. Changes here are frequent, often aligning with new releases or updates. They directly impact how the application operates and serves users. We predominantly use Helm for this layer, focusing on agility and consistency across different environments. This layer reflects the culmination of development efforts, transitioning from testing to real-world application.

Directory Structure

Your directory structure will look like this:

- **00-vpceks**

    Here, we define the virtual private network, elastic Kubernetes cluster, and some foundational DNS settings.

- **01-k8s**

    This is the home for Kubernetes add-ons, ingress, and storage functionality.

- **02-karpenter**

    Karpenter is our cluster autoscaler component that makes our infrastructure highly cost-effective.

- **03-gitlab**

    GitLab community edition cluster.

- **04-webide**

    This is where we set up the Coder environment—an amazing piece of software that provides a fully functional IDE in the browser, running inside our Kubernetes cluster.

A Few Words About Uniqueness

This book is the essence of my experience building and managing hundreds of Kubernetes clusters and handing them over to other professionals. Over the years, I've never seen two identical IT infrastructures. It's amazing—we use the same tools, the same cloud providers, yet everyone develops a unique approach to managing infrastructure. So, use this book to build your own solution. Keep it as a reference point. Consider it a collection of patterns from which you'll select the ones that work best for you.

Terraform 101

The Early Days of Infrastructure

In the era before cloud computing, managing virtual infrastructure on a cluster of bare-metal machines was a complex task. We used Libvirt, virtual ethernet devices, and VLAN networking to handle this. This process, heavily reliant on scripting, involved calling APIs, managing timeouts, and constantly verifying resource states. The imperative nature of this approach often led to high-stress situations, with significant risks associated with errors.

The Declarative Revolution

Terraform marked a significant shift in infrastructure management. Moving away from an imperative to a declarative approach, it allows you to define your infrastructure's desired end state. This method emphasizes planning and thorough validation, streamlining management and reducing error rates.

Declaring a Resource

Declaring a resource in Terraform is straightforward. The syntax is typically as follows:

```
resource "<provider-name>_<kind-of-resource>" "<name>" {
    // Properties
}
```

Please note, this is not what you see once the resource is created on the remote side. Instead, this is the key by which Terraform stores the resource's data in a state file. The provider-kind-name should be unique within your workspace.

State Management

Executing terraform apply initiates a complex interaction between your configuration, the state file, and the cloud provider's API. Securely storing this state, preferably in a cloud service like an S3 bucket, is vital:

```
terraform {
  backend "s3" {
    region = "<your-region>"
    bucket = "<bucket-name>"
    key    = "<unique-key>"
  }
}
```

Providers

Terraform operates with various providers for resource management. For AWS resources, for example, the configuration looks like this:

```
provider "aws" {
    // Provider-specific properties
}
resource "aws_instance" "bastion" {
    // Instance-specific properties
}
```

Initializing Terraform with terraform init is essential to prepare your environment for these configurations.

Fetching Data

Retrieving data from remote sources is a common requirement. For instance, to obtain the ARN of a pre-existing S3 bucket named "foobar," you would use

```
data "aws_s3_bucket" "selected" {
    bucket = "foobar"
}
```

Accessing its properties is as simple as data.aws_s3_bucket.selected.arn.

Local Variables

Terraform's local variables help minimize repetitive typing and reduce typos:

```
locals {
    bucket_arn = data.aws_s3_bucket.selected.arn
}
```

Then, later in the code, you can reference it with local.bucket_arn or ${local.bucket_arn}/. *The last statement with /* means the files in the bucket.

Input Variables

Terraform allows for external variable definitions to enhance configurability:

```
variable "bucket_name" {
    type = string
    description = "Name of the bucket for image storage"
    default = "foobar"
}
```

The usage of these variables will be explored further.

Outputs

Outputting values in Terraform is achieved through

```
output "bucket_arn" {
    value = aws_s3_bucket.selected.arn
}
```

Their practical applications will be discussed in later sections.

Organizing Your Working Directory

Files in your Terraform project should have the .tf extension. While filenames can vary, certain conventions and special filenames exist, which will be discussed in subsequent chapters.

Ready to embark on building your infrastructure? Let's advance to the next chapter.

CHAPTER 2

Starting from Scratch

Beginning of the Journey: VPC, DNS, and ECR

The Current Phase of the Organization

It's crucial that our DevOps strategy aligns seamlessly with the organization's current life stage. In this book, we're focusing on a startup gaining momentum, positioned in a complex and challenging phase. The need to scale is paramount—not just in doubling the developer team, but also in gearing up for a tenfold increase in clients.

DevOps Goals

Our primary goal is a cost-effective, highly dynamic development environment. We anticipate running at least ten versions of the infrastructure concurrently, encompassing demos, open PRs, and staging. This will be a playground for testing, feature discussions, and experimentation, not yet a production environment. Additional objectives include designing the system so that it requires minimal maintenance, ideally zero. Hence, our priorities are security, simplicity, and cost-effectiveness.

CHAPTER 2   STARTING FROM SCRATCH

Minimal Required Work via Web Console

We begin by logging into the AWS console. Once there, decide on the region—typically us-east-1 for US projects and eu-west-1 for EU ones. Next, create a bucket for our Terraform state in S3, ensuring the name is globally unique. Enable versioning but keep other defaults.

For Terraform scripts, create a user with full admin privileges. Ideally, a role is preferable, but that adds complexity, which we want to avoid in this book. Once the user is created, generate and download the access keys.

Setting Up CLI

Now, secure your keys. Create a file with the following content:

```
export AWS_DEFAULT_REGION=<region>
export AWS_ACCESS_KEY_ID=<key-id>
export AWS_SECRET_ACCESS_KEY=<key-secret>
```

Source this file before starting your Terraform journey. Ensure this path is in your .gitignore—never store keys or secrets in git!

Windows users, use the following commands:

```
$env:AWS_DEFAULT_REGION = "<region>"
$env:AWS_ACCESS_KEY_ID = "<key-id>"
$env:AWS_SECRET_ACCESS_KEY = "<key-secret>"
```

or

```
set AWS_DEFAULT_REGION=<region>
set AWS_ACCESS_KEY_ID=<key-id>
set AWS_SECRET_ACCESS_KEY=<key-secret>
```

Confirm setup with `aws sts get-caller-identity`.

If you encounter issues installing awscli, here's my preferred method:

```
python3 -m venv .venv
source .venv/bin/activate # on Windows .venv\Scripts\activate
pip install awscli
```

Creating State Bucket via CLI

You might want to create the bucket for storing Terraform state via CLI instead of using a web console. Here is the way:

```
aws s3api create-bucket --bucket your-bucket-name --region your-region

aws s3api put-bucket-versioning --bucket your-bucket-name --versioning-configuration Status=Enabled
```

First Steps in Our Infra Git Repo

Begin by creating our first directory and initializing Terraform files:

```
mkdir 00-vpceks
cd 00-vpceks
```

Despite a decade each with Emacs and Vim, I am guilty of using VSCode. ;) If we're on the same page, continue with code.

Variables in variables.tf

```
// variables.tf
variable "vpc_cidr" {
    default = "10.18.0.0/16"
}
```

CHAPTER 2 STARTING FROM SCRATCH

```
variable "tag" {
    default = "<your-tag>"
}
variable "region" {
    default = "us-east-1" //put yours here
}
```

In Terraform, define the IP range for your VPC and a tag for naming and tagging your resources. Ensure `vpc_cidr` doesn't overlap with other VPCs' IPs.

Tag is supposed to be the name or in the name of resources we are creating. I usually choose something related to a domain name I use for this environment. If the domain is foobar.com the tag might be foobar-com. Or whatever you pleased. But please use characters, numbers, _, and -.

Main.tf

Define where to store state, set the Terraform provider, and establish some local variables for efficiency.

```
// main.tf
terraform {
    backend "s3" {
    bucket = "<bucket-name>"
    key = "00-vpceks.tfstate"
    region = "<region>"   // we could not use var here
    }
}
provider "aws" {
    region = var.region   // but here we can
}
data "aws_caller_identity" "current" {}
```

```
data "aws_availability_zones" "zones" {}
```

```
locals {
    account_id = data.aws_caller_identity.current.account_id
}
```

After creating the files, run `terraform init`.

Building the VPC (Virtual Private Cloud)

Let's create `vpc.tf` with the following content:

```showLineNumbers
module "vpc" {
    source = "terraform-aws-modules/vpc/aws"
    // check the current version at:
    // https://registry.terraform.io/modules/terraform-aws-modules/vpc/aws/latest
    version = "5.13.0"

    name = var.tag
    cidr = var.vpc_cidr
    azs  = slice(data.aws_availability_zones.zones.names, 0, 3)

    private_subnets = [
        cidrsubnet(var.vpc_cidr, 8, 1),
        cidrsubnet(var.vpc_cidr, 8, 2),
        cidrsubnet(var.vpc_cidr, 8, 3)
    ]
    public_subnets = [
        cidrsubnet(var.vpc_cidr, 8, 4),
        cidrsubnet(var.vpc_cidr, 8, 5),
        cidrsubnet(var.vpc_cidr, 8, 6)
    ]
```

CHAPTER 2   STARTING FROM SCRATCH

```
    enable_nat_gateway = true
    single_nat_gateway = true
    one_nat_gateway_per_az = false

  private_subnet_tags = {
    "karpenter.sh/discovery"           = var.tag
    "kubernetes.io/role/internal-elb"  = "1"
    "kubernetes.io/cluster/${var.tag}" = "shared"
  }
  public_subnet_tags = {
    "kubernetes.io/role/elb"           = "1"
    "kubernetes.io/cluster/${var.tag}" = "shared"
  }
}

resource "aws_security_group_rule" "allow_outbound" {
    description = "Outbound access"
    protocol = "tcp"
    type = "egress"
    from_port = 0
    to_port = 65535
    cidr_blocks = ["0.0.0.0/0"]
    security_group_id = module.vpc.default_security_group_id
}

resource "aws_security_group_rule" "vpc_access" {
    description = "Inside VPC"
    protocol = "tcp"
    type = "ingress"
    from_port = 0
    to_port = 65535
    cidr_blocks = [var.vpc_cidr]
    security_group_id = module.vpc.default_security_group_id
}
```

In this section, `module vpc` indicates that we are using a Terraform module. We source this module from `registry.terraform.io`. The `azs` slice specifies that we spread our networking across three availability zones, which becomes significant when working with volumes.

This module sets up a VPC with six subnets, three public and three private. The public subnets are connected directly to the Internet gateway and can be accessed publicly. Instances in the private subnets communicate with the internet via a NAT gateway. We've opted for a single NAT gateway for all subnets. Note that NAT gateways incur costs.

The `cidrsubnet` function generates specific subnet ranges within our VPC CIDR block. For example, `cidrsubnet(var.vpc_cidr, 8, 1)` gives us `10.18.1.0/24` from `10.18.0.0/16`. The `private_subnet_tags` will be discussed in a later chapter.

The `resource "aws_security_group_rule"` entries define the default security group rules for our instances, allowing outbound access to anywhere and inbound access from within the VPC.

After defining your VPC in `vpc.tf`, run `terraform init` to integrate the new module, followed by `terraform apply` to create your VPC infrastructure. Enjoy the process as you witness the construction of your VPC.

Dealing with DNS (Domain Name System)

Typically, domain registration is straightforward. In this book, we are going to use a domain for the whole environment. I expect you have it registered at some registrar, like GoDaddy, but managing records we will be doing from Terraform.

Now, add a variable in `variables.tf`:

```
//variables.tf
variable "domain" {
    default = "<your-domain>"
}
```

CHAPTER 2   STARTING FROM SCRATCH

Next, create dns.tf with the following content:

```
resource "aws_route53_zone" "this" {
  name = var.domain
}
resource "aws_route53_record" "validation" {
  for_each = {
    for dvo in aws_acm_certificate.this.domain_validation_
    options : dvo.domain_name => {
      name   = dvo.resource_record_name
      record = dvo.resource_record_value
      type   = dvo.resource_record_type
    }
  }

  allow_overwrite = true
  name            = each.value.name
  records         = [each.value.record]
  ttl             = 60
  type            = each.value.type
  zone_id         = aws_route53_zone.this.zone_id
}
resource "aws_acm_certificate_validation" "this" {
  certificate_arn         = aws_acm_certificate.this.arn
  validation_record_fqdns = [for record in aws_route53_record.
                            validation : record.fqdn]
}
```

This way we do create a Route53 hosting zone to help us manage the DNS records. Plus we do create SSL certificates.

Let us have some DNS-related outputs. Please create `outputs.tf` in the very same folder and put the following:

```
output "nameservers" {
    value = aws_route53_zone.this.name_servers
}
```

ECR (Elastic Container Registry)

What I appreciate most about Terraform is its self-explanatory nature. You can simply look at the code and understand its purpose:

```
resource "aws_ecr_repository" "web" {
  name                 = "${var.tag}-web"
  image_tag_mutability = "MUTABLE"
  image_scanning_configuration {
    scan_on_push = true
  }
}

resource "aws_ecr_lifecycle_policy" "this" {
  repository = aws_ecr_repository.web.name

  policy = jsonencode({
    rules = [
      {
        rulePriority = 1
        description = "Expire images older than the 5 most recent"
        selection = {
          tagStatus = "any"
          countType = "imageCountMoreThan"
          countNumber = 5
        }
        action = {
          type = "expire"
```

```
      }
    }
  ]
})
}
resource "aws_vpc_endpoint" "ecr-dkr-endpoint" {
  vpc_id               = module.vpc.vpc_id
  private_dns_enabled  = true
  service_name         = "com.amazonaws.${var.region}.ecr.dkr"
  vpc_endpoint_type    = "Interface"
  security_group_ids   = [module.vpc.default_security_group_id]
  subnet_ids           = module.vpc.private_subnets
}
```

Here, we create a repository for our Docker images, enforce a security scan with every upload, and maintain only the five latest images. Additionally, we set up an endpoint to economize our usage, bypassing the NAT gateways for downloading images. I expect you to put this content into ecr.tf.

Please also update your outputs.tf with the following:

```
output "ecr_url" {
  value = aws_ecr_repository.web.repository_url
}
```

Now let's talk about permissions. Turn the page.

CHAPTER 3

Building the Infrastructure Base

Our First IAM Role: AWS Identity and Access Management 101

How I Expected It to Be

I spent way too much time and effort trying to wrap my mind around AWS IAM. I started with the wrong expectations. In this book, I'm going to simplify things for you and help you avoid the same mistakes I made. I initially thought IAM would be similar to real life, where you have an organization with a set of roles, and each role has a set of permissions. For example, in an organization, a CTO might have access to a bank account. This approach is known as RBAC—Role-Based Access Control. By the way, Kubernetes uses this approach. But not AWS!

AWS IAM Is Not RBAC

First, understand this: you can start working with AWS without ever creating a role. What you need is a *policy*. The simplest and most powerful policy looks like this:

```
{
    "Version": "2012-10-17",
    "Statement": [
        {
            "Effect": "Allow",
            "Action": "*",
            "Resource": "*"
        }
    ]
}
```

This policy lets you do anything with everything. But who is "you"? Well, "you" could be a user or a user group. And how do we attach the policy? There are two ways: inline policies or managed policies. See the docs for details, but we usually go with managed policies. Are there other objects where we can attach a policy? For instance, what if we need a service to get permissions? The answer is yes—we can use a role.

The Role of a Role

A role is an object where we can attach a policy, and then, this role can be assumed by an AWS service or user. For example, if we need a service to read from our S3 bucket named "foobar," we would first create a policy like this:

```
    "Version": "2012-10-17",
    "Statement": [
        {
            "Effect": "Allow",
            "Action": [
                "s3:GetObject",
                "s3:ListBucket"
            ],
```

```
            "Resource": [
                "arn:aws:s3:::foobar",
                "arn:aws:s3:::foobar/*"
            ]
        }
    ]
}
```

Then, we would create a role—let's call it ReadFooBar and attach that policy to the role and, finally, allow the AWS service to assume this role. The way we accomplish this last step depends on the service itself. You'll see specific examples later in the book.

We Need an IAM Policy for Our Valet

Remember the main topic of this book? Zero-maintenance strategies. This means zero work for us, but something else has to do our work. We need a valet. We're going to use an amazing piece of software for that—Karpenter.sh. It will handle scaling the cluster up and down for us, which means it needs to be able to create and delete virtual machines and do a lot of other stuff. So, let's create a policy for its job.

Karpenter Policy

While you're in the 00-vpceks folder, create a file named karpenter.tf. Remember, you can use any filename as long as it has a .tf extension.

```
resource "aws_iam_policy" "karpenter" {
  name   = "KarpenterPolicy-${var.tag}"
  path   = "/"
  policy = <<EOF
{
```

```json
  "Version": "2012-10-17",
  "Statement": [
    {
      "Action": [
        "ssm:GetParameter",
        "ec2:*",
        "pricing:GetProducts"
      ],
      "Effect": "Allow",
      "Resource": "*",
      "Sid": "Karpenter"
    },
    {
      "Action": "ec2:TerminateInstances",
      "Condition": {
        "StringLike": {
          "ec2:ResourceTag/karpenter.sh/nodepool": "*"
        }
      },
      "Effect": "Allow",
      "Resource": "*",
      "Sid": "ConditionalEC2Termination"
    },
    {
      "Effect": "Allow",
      "Action": "iam:PassRole",
      "Resource": "*",
      "Sid": "PassNodeIAMRole"
    },
    {
      "Effect": "Allow",
```

```
    "Action": "eks:DescribeCluster",
    "Resource": "arn:aws:eks:${var.region}:${local.account_
               id}:cluster/${var.tag}",
    "Sid": "EKSClusterEndpointLookup"
  },
  {
    "Sid": "AllowScopedInstanceProfileCreationActions",
    "Effect": "Allow",
    "Resource": "*",
    "Action": [
      "iam:CreateInstanceProfile"
    ],
    "Condition": {
      "StringEquals": {
        "aws:RequestTag/kubernetes.io/cluster/${var.tag}":
        "owned",
        "aws:RequestTag/topology.kubernetes.io/region":
        "${var.region}"
      },
      "StringLike": {
        "aws:RequestTag/karpenter.k8s.aws/ec2nodeclass": "*"
      }
    }
  },
  {
    "Sid": "AllowScopedInstanceProfileTagActions",
    "Effect": "Allow",
    "Resource": "*",
    "Action": [
      "iam:TagInstanceProfile"
    ],
```

```
      "Condition": {
        "StringEquals": {
          "aws:ResourceTag/kubernetes.io/cluster/${var.tag}":
          "owned",
          "aws:ResourceTag/topology.kubernetes.io/region":
          "${var.region}",
          "aws:RequestTag/kubernetes.io/cluster/${var.tag}":
          "owned",
          "aws:RequestTag/topology.kubernetes.io/region":
          "${var.region}"
        },
        "StringLike": {
          "aws:ResourceTag/karpenter.k8s.aws/ec2nodeclass": "*",
          "aws:RequestTag/karpenter.k8s.aws/ec2nodeclass": "*"
        }
      }
    },
    {
      "Sid": "AllowScopedInstanceProfileActions",
      "Effect": "Allow",
      "Resource": "*",
      "Action": [
        "iam:AddRoleToInstanceProfile",
        "iam:RemoveRoleFromInstanceProfile",
        "iam:DeleteInstanceProfile"
      ],
      "Condition": {
        "StringEquals": {
          "aws:ResourceTag/kubernetes.io/cluster/${var.tag}":
          "owned",
          "aws:ResourceTag/topology.kubernetes.io/region":
          "${var.region}"
```

```
      }
    }
  },
  {
    "Sid": "AllowInstanceProfileReadActions",
    "Effect": "Allow",
    "Resource": "*",
    "Action": "iam:GetInstanceProfile"
  },
  {
    "Sid": "webshield",
    "Effect": "Allow",
    "Action": [
      "wafv2:*",
      "shield:*",
      "waf-regional:*"
    ],
    "Resource": "*"
  },
  {
    "Sid": "SQS",
    "Effect": "Allow",
    "Action": "sqs:*",
    "Resource": "${aws_sqs_queue.karpenter_interruption_queue.arn}"
  },
  {
    "Effect": "Allow",
    "Action": [
      "iam:CreateServiceLinkedRole"
    ],
```

```
      "Resource": "arn:aws:iam::*:role/aws-service-role/spot.
      amazonaws.com/AWSServiceRoleForEC2Spot",
      "Condition": {
        "StringEquals": {
          "iam:AWSServiceName": "spot.amazonaws.com"
        }
      }
    }
  ]
}
EOF
}
```

It's a lot to digest, I know. I pulled it straight from the Karpenter documentation and adapted it a bit to work with our Terraform setup. Honestly, I don't understand every single detail, but you know the golden rule of DevOps: if it's not broken, don't fix it.

Dramatically Reducing Compute Bills

Karpenter offers significant cost savings, especially when configured for SPOT instances. SPOT instances are much cheaper but can be terminated at any time. Karpenter mitigates this risk by pre-emptively launching new instances, transferring workloads, and gracefully shutting down the old ones. To enable this smart handling, we need some additional infrastructure in place:

```
resource "aws_sqs_queue" "karpenter_interruption_queue" {
  name                      = "${var.cluster_name}-queue"
  message_retention_seconds = 300
  kms_master_key_id         = "alias/aws/sqs"
}
```

```
resource "aws_sqs_queue_policy" "karpenter_interruption_queue_
policy" {
  queue_url = aws_sqs_queue.karpenter_interruption_queue.id

  policy = jsonencode({
    Id      = "EC2InterruptionPolicy",
    Version = "2008-10-17",
    Statement = [
      {
        Effect = "Allow",
        Principal = {
          Service = [
            "events.amazonaws.com",
            "sqs.amazonaws.com"
          ]
        },
        Action   = "sqs:SendMessage",
        Resource = aws_sqs_queue.karpenter_interruption_
                   queue.arn
      }
    ]
  })
}

resource "aws_cloudwatch_event_rule" "scheduled_change_rule" {
  name = "ScheduledChangeRule"
  event_pattern = jsonencode({
    source = [
      "aws.health"
    ],
    "detail-type" = [
      "AWS Health Event"
```

CHAPTER 3   BUILDING THE INFRASTRUCTURE BASE

```
    ]
  })
}
resource "aws_cloudwatch_event_target" "scheduled_change_
target" {
  rule      = aws_cloudwatch_event_rule.scheduled_change_
              rule.name
  target_id = "KarpenterInterruptionQueueTarget"
  arn       = aws_sqs_queue.karpenter_interruption_queue.arn
}
```

In this configuration, we create an SQS queue to receive health event notifications. EventBridge then routes these events to the queue, enabling Karpenter to respond promptly to instance interruptions, ensuring a smooth and cost-effective operation.

CHAPTER 4

Sculpting Kubernetes: The Initial Steps

Introducing Kubernetes

How Would You Do That?

Okay, imagine you are the CTO of a big IT company providing a platform as a service. So you have hundreds of thousands of applications running at the same time. Applications of very different kinds. Where would you start? Well, if you are like me, you would start with the database. Kubernetes team chose etcd for that role. Let me explain why.

Kubernetes Database: Etcd

In the world of distributed systems, chaos lurks at every corner. You need a way to bring order to that chaos, something rock solid. Enter etcd—the unsung hero that makes Kubernetes tick. Think of it like the heartbeat that keeps the system alive. Without etcd, Kubernetes would be like a car without an engine.

Now, you might ask, why etcd? What makes it so special? Let's break it down.

Reliability

etcd doesn't flinch under pressure. It's built for fault tolerance. When the going gets tough, etcd keeps your data consistent across the board, even if parts of your system fail. In the cutthroat world of distributed systems, reliability isn't just important—it's everything.

Consistency

In a world where decisions need to be made fast, consistency is non-negotiable. etcd uses the Raft consensus algorithm to ensure that when you ask for data, you get the right data—every single time. No guessing, no second chances. It's a deal you can trust.

Speed

etcd doesn't just sit around waiting for things to happen. It's quick, with fast reads and writes, allowing Kubernetes to stay responsive. Speed is crucial, especially when every millisecond counts.

Scalability

You don't want a system that crumbles as you grow. etcd scales with you, handling thousands of clients without breaking a sweat. It's built for the big leagues.

Simplicity

Complexity kills. etcd's interface is simple, so developers can integrate it without jumping through hoops. This simplicity is a major reason why Kubernetes chose etcd as its backbone.

Watch: Real-Time Monitoring

etcd's watch is like having a real-time monitoring system built into your infrastructure. Whenever there's a change—whether it's an update to configuration data or a new node joining the cluster—etcd instantly notifies Kubernetes. There's no delay, no need for constant polling. It's as if every change in your infrastructure sends a direct alert, allowing Kubernetes to respond immediately.

Why This Is Crucial

For sys admins, this is critical. In large-scale environments, things change rapidly. You need a system that can keep up. With etcd's watch, Kubernetes can automatically adapt to changes in the environment as soon as they happen. Whether it's scaling applications, rerouting traffic, or deploying updates, etcd ensures Kubernetes is always in sync with the current state of the system.

A Control Hub

Well, we have a database, we can watch for changes and now what? How we would control everything? Enter API Server.

The Kubernetes API Server

Think of the Kubernetes API Server as the central nervous system of your Kubernetes cluster. It's the control hub where everything comes together. Every command, every query, every change you want to make in your cluster—it all flows through the API Server.

CHAPTER 4   SCULPTING KUBERNETES: THE INITIAL STEPS

Single Point of Control

As a sys admin, you know the importance of having a single point of control. The Kubernetes API Server provides exactly that. Whether you're deploying applications, scaling services, or monitoring the health of your system, you interact with the API Server. It takes your instructions and ensures they're executed across the entire cluster, no matter how large or complex.

RESTful Interface

The beauty of the API Server lies in its simplicity and power. It's built on a RESTful interface, meaning you can interact with it using standard HTTP methods. This makes it incredibly flexible—whether you're using kubectl, integrating with CI/CD pipelines, or developing custom automation scripts, the API Server is the gateway that makes it all possible.

Authentication and Security

But it's not just about control; it's about secure control. The Kubernetes API Server is also the gatekeeper. It manages authentication and authorization, ensuring that only the right people and processes have access to the cluster. You can define who can do what, down to the finest details. This is crucial for maintaining security and compliance in a multi-tenant environment.

Real-Time State Management

One of the most powerful aspects of the API Server is how it manages the state of your cluster. It's constantly communicating with etcd to keep a real-time record of the desired state versus the actual state. If there's a discrepancy—say a pod crashes or a node goes offline—the API Server is immediately aware and can initiate corrective actions. This ensures that your cluster remains stable, even as conditions change.

Extensibility

Finally, the API Server isn't just a static piece of infrastructure. It's extensible. You can add custom resources and extend Kubernetes to meet your specific needs. This allows Kubernetes to grow with your environment, supporting new workflows and technologies as they emerge.

Summary

In summary, the Kubernetes API Server is the nerve center of your cluster—handling communication, control, security, and real-time state management. For sys admins, it's the indispensable tool that makes managing complex, dynamic environments not just possible, but efficient and secure. Without it, Kubernetes simply wouldn't function.

It's a Lot of Servers

Just imagine how many servers are needed to manage OpenAI infrastructure. Some of them with GPU others with a lot of RAM but CPU only. We need the brain that decides where applications run within the cluster.

The Kubernetes Scheduler: The Decision-Maker

In a Kubernetes cluster, the Scheduler is like a master air traffic controller. It's the component that decides which node in your cluster will run a given workload. When you deploy an application, the Scheduler is the one making the critical decisions about where that application will live, based on the current state of the cluster.

Matching Workloads to Resources

Imagine you've got a dozen planes coming in to land, and you need to figure out which runway each one should use. That's what the Scheduler does with your pods. It looks at the resource requirements of each pod—CPU, memory, storage—and matches them to the available resources on the nodes in your cluster. The Scheduler ensures that each workload lands on a node that can handle it efficiently.

Optimizing Performance

But the Scheduler doesn't just assign workloads randomly. It's designed to optimize the overall performance of the cluster. It considers a variety of factors like resource requests, node capacity, and even specific policies you've defined. The goal is to maximize resource utilization while maintaining balance across the nodes. This prevents any single node from being overwhelmed while others sit idle.

Affinity and Anti-affinity

For sys admins managing complex environments, control over placement is crucial. The Scheduler supports affinity and anti-affinity rules, allowing you to dictate how pods should be distributed. Need to keep certain workloads together for performance reasons? Use affinity. Want to spread critical applications across different nodes for high availability? Anti-affinity has you covered. The Scheduler respects these rules, giving you fine-grained control over how your workloads are deployed.

Handling Constraints

The Scheduler is also adept at handling constraints. Whether you have specific hardware requirements, need to avoid certain nodes, or need to comply with security and compliance regulations, the Scheduler factors in all these constraints. It ensures that your workloads are placed in a way that meets your operational requirements.

Real-Time Decision-Making

What sets the Scheduler apart is its ability to make these decisions in real time. As the state of your cluster changes—nodes coming online or going offline, workloads scaling up or down—the Scheduler constantly recalculates the best placement for new and existing pods. It's this dynamic decision-making capability that keeps your cluster running smoothly, even as conditions change.

Extensibility

Just like other components in Kubernetes, the Scheduler is extensible. You can customize its behavior with custom scheduling policies or even create your own Scheduler for specific workloads. This flexibility ensures that the Scheduler can adapt to your unique environment and operational needs.

Summary

In short, the Kubernetes Scheduler is the smart decision-maker of your cluster. It's the one ensuring that your workloads are placed efficiently, balancing performance, and adhering to your policies and constraints. For sys admins, the Scheduler is the key to maintaining a well-organized, efficient, and responsive cluster. Without it, Kubernetes would be chaotic and unpredictable, instead of the reliable orchestration platform it is.

One to Control Everything Kubernetes Controller Manager

Let's dig into the **Kubernetes Controller Manager**, one of the unsung heroes of the Kubernetes control plane.

CHAPTER 4   SCULPTING KUBERNETES: THE INITIAL STEPS

Kubernetes Controller Manager: The Orchestrator

The Kubernetes Controller Manager is like the orchestra conductor for your cluster. It's responsible for ensuring that the desired state of your system, as declared by you, matches the actual state in your cluster. If there's a discrepancy, the Controller Manager takes action to bring everything back in line.

What Is a Controller?

Before we dive deeper, let's talk about **controllers**. In Kubernetes, a controller is a loop that watches the state of your cluster, and when it detects that the actual state doesn't match the desired state, it makes changes to reconcile them. Controllers are the brains behind maintaining, scaling, and healing your cluster.

Key Controllers Managed by the Controller Manager

1. **Replication Controller**: The **Replication Controller** ensures that a specified number of pod replicas are running at any given time. If you say you want three replicas of a pod, the Replication Controller makes sure that exactly three are running—no more, no less. If one pod crashes, the Replication Controller will spin up another to replace it.

2. **Node Controller**: The **Node Controller** is responsible for monitoring the health of nodes. It detects if a node goes down and takes corrective action. This might involve marking the node as unavailable and scheduling pods that were running on that node onto other healthy nodes. This ensures your workloads remain up and running even if some of your hardware fails.

3. **Job Controller**: The **Job Controller** manages batch jobs, ensuring that a specified number of pods complete their tasks. It's perfect for workloads that are not continuous but need to run to completion, like data processing tasks or backups.

4. **Endpoint Controller**: The **Endpoint Controller** manages the association between services and pods. It ensures that the service is always pointing to the correct set of pods, so your applications can reliably communicate with each other, even as the underlying pods scale up or down.

5. **Service Account and Token Controllers**: These controllers create default service accounts and API access tokens for new namespaces, ensuring that your pods can securely communicate with the Kubernetes API Server.

6. **DaemonSet Controller**: The **DaemonSet Controller** ensures that a copy of a pod runs on all (or a specific subset of) nodes. This is particularly useful for running system-level pods like logging or monitoring agents on every node in your cluster.

How It Works

The Controller Manager runs as a single process but manages multiple controllers. Each of these controllers is responsible for one or more resources in your cluster, such as pods, services, endpoints, and nodes. The Controller Manager constantly watches the state of these resources via the Kubernetes API, and whenever it detects that something isn't as it should be, it takes action to correct it.

For example, if a node fails, the Node Controller will detect that the pods on that node are no longer available and will work with the Scheduler to relocate those pods to a healthy node. Similarly, if the number of replicas for a deployment drops below the desired count, the Replication Controller will create new pods to bring the count back up.

The Heartbeat of Kubernetes

The Controller Manager is essentially the heartbeat of Kubernetes, constantly working behind the scenes to keep your cluster in the desired state. Without it, your cluster would be static and unresponsive to changes. For sys admins, the Controller Manager provides peace of mind—knowing that Kubernetes is continuously monitoring the cluster and making sure everything runs as expected.

Extensibility

The Controller Manager is also extensible. Kubernetes allows you to create custom controllers to manage your own resources or extend the behavior of existing ones. This flexibility is part of what makes Kubernetes so powerful and adaptable to different workloads and environments.

Summary

In short, the Kubernetes Controller Manager is the orchestrator that ensures everything in your cluster is working as it should. It's the component that keeps your environment self-healing, scalable, and robust, automating the tedious aspects of cluster management so you can focus on what really matters.

CHAPTER 4   SCULPTING KUBERNETES: THE INITIAL STEPS

Where Are the Real Workers? Master vs. Worker Nodes

Let's break down what's happening on a **Kubernetes node**—the workhorse of your cluster.

Kubernetes Node: The Workhorse

A Kubernetes node is where the rubber meets the road. It's the actual machine—physical or virtual—where your containers run. When you deploy an application in Kubernetes, it's the nodes that do the heavy lifting, running your workloads and keeping everything operational.

Key Components on a Kubernetes Node

1. **Kubelet**: The **kubelet** is the agent on each node that makes sure containers are running as they should be. It's constantly communicating with the Kubernetes API Server, receiving instructions, and reporting the status of the node and its pods. The kubelet is responsible for starting, stopping, and maintaining the containers based on the PodSpecs defined by the cluster.

2. **Container Runtime**: The **container runtime** is the software that actually runs the containers. Kubernetes supports several runtimes, with Docker and containerd being the most common. This component pulls the necessary container images from a registry and starts the containers as defined by the kubelet. Without the container runtime, there's no way to run your containers on the node.

3. **Kube-proxy**: The **kube-proxy** handles networking on the node. It ensures that each pod gets its own unique IP address and manages network routing for the pods. It also handles load balancing for services and ensures that network requests are correctly routed to the appropriate pod, whether it's on the same node or across the cluster.

4. **Pod**: The **pod** is the smallest deployable unit in Kubernetes. A pod can contain one or more containers that share the same network namespace, IP address, and storage. When you deploy an application, it's deployed as one or more pods across the nodes in the cluster. Each pod is isolated from others and has its own environment, making it an ideal way to run microservices or other containerized applications.

5. **cAdvisor**: **cAdvisor** is an agent that runs on each node to monitor the resource usage and performance of the containers. It collects and provides metrics like CPU, memory, network, and filesystem usage, which can be used by the kubelet and other tools for monitoring and scaling purposes.

6. **Node Components for Storage**: Kubernetes nodes also handle persistent storage for your pods. When a pod requires storage that persists beyond its lifecycle, the node connects to storage solutions like NFS, iSCSI, or cloud-based storage through Persistent Volumes (PVs) and Persistent Volume Claims (PVCs). The kubelet manages these storage volumes, ensuring that they're properly mounted and accessible to the pods.

CHAPTER 4   SCULPTING KUBERNETES: THE INITIAL STEPS

7. **Operating System**: Beneath all these components is the **operating system** of the node, usually a Linux distribution. The OS is tuned and configured to support containerized workloads efficiently, providing the necessary resources, security, and networking capabilities that Kubernetes needs to operate effectively.

Node Roles: Worker vs. Master

While every node runs these core components, there's a distinction between **worker nodes** and **master nodes** (control plane nodes). Worker nodes run your application workloads, while master nodes host the Kubernetes control plane components (API Server, etcd, Scheduler, Controller Manager). However, in smaller clusters, master nodes can also run workloads, but it's common practice to keep these roles separate for stability and security.

The Big Picture

In essence, a Kubernetes node is a self-sufficient unit that contains everything needed to run and manage containers. It's where your applications live and breathe. The components on the node work together to ensure that containers are running smoothly, are properly networked, and have access to the resources they need.

Well, it's time to talk about AWS. We continue with the very complicated topic Identity Access Management on AWS. But I believe in use, we will handle it together.

41

CHAPTER 5

Bridging Kubernetes and AWS Resources

Elastic Kubernetes at Your Service

The Standard Kubernetes Cluster

In a typical Kubernetes setup, you start with a standard configuration: three master nodes and multiple worker nodes. This setup provides a balance of high availability and resource distribution, ensuring your cluster is resilient and scalable. However, managing these nodes, especially in a cloud environment like AWS, can become complex and resource-intensive.

Setting Up a Kubernetes Cluster in AWS

When it comes to setting up a Kubernetes cluster in AWS, you have several options:

1. **Kubeadm:** The official tool for bootstrapping a Kubernetes cluster, kubeadm provides a straightforward way to set up a cluster on any infrastructure, including AWS. However, it requires manual management of both master and worker nodes, which can be time-consuming.

2. **Kops**: Kubernetes operations (kops) is a popular choice for creating, upgrading, and managing production-grade Kubernetes clusters on AWS. It automates much of the setup process and integrates well with AWS services, but you still manage the master nodes yourself.

3. **Amazon EKS (Elastic Kubernetes Service)**: A fully managed Kubernetes service provided by AWS, EKS takes care of the master node management, including updates, scaling, and high availability. This significantly reduces the operational overhead, allowing you to focus on deploying and managing your applications on worker nodes.

I've tried all of these options, but once I switched to EKS, I never looked back. EKS simplifies the process, offloading much of the heavy lifting, especially when it comes to managing the master nodes.

Our Choice: EKS

With EKS, AWS manages the master nodes for you, taking care of updates, scaling, and availability, so you can focus on creating and managing the worker nodes. This separation allows you to streamline operations and concentrate on the aspects of your infrastructure that matter most to your applications. EKS integrates seamlessly with other AWS services, making it the ideal choice for a robust, scalable Kubernetes environment.

Our Choice: EKS by Terraform

When it comes to setting up and managing EKS clusters, I chose Terraform over tools like eksctl for several key reasons. Terraform's declarative approach aligns perfectly with GitOps principles, allowing us to define our entire infrastructure as code and version it in Git. This makes the infrastructure reproducible, auditable, and easy to manage over time. By using Terraform, we can ensure that every change is reviewed, tested, and approved before being applied, providing a clear, consistent, and automated workflow. Additionally, Terraform's extensive ecosystem of providers and modules makes it easy to integrate with other AWS services and maintain a unified, scalable infrastructure across our entire cloud environment.

Our Choice: EKS by Terraform EKS Module

Even with Terraform, setting up and managing an EKS cluster can seem daunting, especially when it comes to handing over the infrastructure to another DevOps engineer. What I needed was a way to simplify the process and make it as straightforward as possible—ideally, something that allows me to describe everything I need from the cluster on a single page. Thanks to AWS Community Hero Anton Babenko, we have exactly that in the form of the Terraform EKS Module. This module abstracts much of the complexity, providing a well-documented, reusable, and community-supported solution that encapsulates best practices. It not only simplifies the setup but also ensures consistency and ease of maintenance, making it an ideal choice for teams that value efficiency and reliability in their infrastructure management.

Get to Work

Let's look at the `eks.tf` file, which sets up our EKS cluster using a Terraform module:

```
module "eks" {
  source  = "terraform-aws-modules/eks/aws"
  version = "19.17.4"
```

```
cluster_name      = var.tag
cluster_version   = "1.30"
subnet_ids        = module.vpc.private_subnets
vpc_id            = module.vpc.vpc_id

cluster_endpoint_public_access = true
kms_key_administrators            = [data.aws_caller_identity.
                                     current.arn]

manage_aws_auth_configmap = false
create_aws_auth_configmap = false

cluster_security_group_tags = {
  "karpenter.sh/discovery" = var.tag
}
node_security_group_tags = {
  "karpenter.sh/discovery" = var.tag
}

eks_managed_node_groups = {
  karpenter-30 = {
    min_size       = 2
    max_size       = 2
    desired_size   = 2
    instance_types = ["t3.medium"]
    labels = {
      role = "karpenter"
    }
    iam_role_additional_policies = {
      AmazonSSMManagedInstanceCore = "arn:aws:iam::aws:policy/
      AmazonSSMManagedInstanceCore"
      Karpenter                    = aws_iam_policy.
                                     karpenter.arn
```

```
            VolumeAccess              = "arn:aws:iam::aws:policy/
                                         service-role/
                                         AmazonEBSCSIDriverPolicy"
            LoadBalancer              = "arn:aws:iam::aws:policy/
                                         ElasticLoadBalancingF
                                         ullAccess"
          }
        }
      }
    }
```

Here's a breakdown of the key lines:

- **L1-L3**: We introduce a new module from the Terraform registry with a fixed version for stability.

- **L6**: Specifies the Kubernetes version.

- **L7-L8**: Places the cluster within private subnets, enhancing security.

- **L10**: Enables kubectl and helm to operate from outside our VPC.

- **L11**: Defines who can manage the cluster's KMS encryption keys.

- **L13-L14**: Management of the aws-auth ConfigMap is delegated to another module or script.

- **L16-L21**: These tags are necessary for Karpenter to discover and manage resources, which we'll dive into later.

CHAPTER 5   BRIDGING KUBERNETES AND AWS RESOURCES

- **L23-L36**: Defines a managed node group specifically for Karpenter, set to run on two t3.medium instances. This ensures our system services, including Karpenter, operate reliably, while Karpenter dynamically manages the other nodes in the cluster.

- **L31**: Ensures connectivity to instance shells from the AWS Management Console.

- **L32**: Grants the service's pods on these nodes the ability to manage AWS resources

Summary of Level 0

This concludes our setup for Level 0—the foundational layer that we'll touch maybe once a year, if that. Sure, we'll add small adjustments here and there during the installation phase, but the heavy lifting is done. See you in the next chapter, where we'll dive into a more dynamic layer of our infrastructure.

It's Time to Make It Alive

```
cd 00-vpceks
terraform init
terraform apply
```

It will take around 15 minutes to set up everything. Enjoy.

Custom Nameservers for Domain

In terraform apply output, you see a list of nameservers. You need to use this data to set custom nameservers at your domain registrar. The exact way is vary from registrar to registrar. Please look into the registrar docs for how to.

CHAPTER 6

The Structure of Application

Common Kubernetes Resources Part 1

Introduction Level 1

Well, on this level, we are setting up things we are not going to change really frequently. Maybe we would change some versions every quarter or so. So, let's create a folder 01-k8s.

Variables

Put 01-k8s/variable.tf with the following content:

```
variable "karpenter_version" {
  default = "1.0.0"
}
variable "cluster_name" {
  default = "<name>"    // it should be the value of variable tag from 00-vpceks
}
```

CHAPTER 6   THE STRUCTURE OF APPLICATION

```
variable "karpenter_master_node_label_name" {
  default = "role"
}
variable "karpenter_master_node_label_value" {
  default = "karpenter"
}
variable "storage_class" {
  default = "ebs"
}
```

These variables allow you to customize the deployment of resources in this module. For instance, karpenter_version sets the version of Karpenter, and cluster_name specifies the name of your EKS cluster. The karpenter_master_node_label_name and karpenter_master_node_label_value variables define labels for identifying nodes managed by Karpenter.

Main.tf with State and Providers

Navigate to the 01-k8s directory and create a file named main.tf with the following content:

```
terraform {
    backend "s3" {
    bucket = "<bucket-name>"
    key = "01-k8s.tfstate"
    region = "<region>"
    }
}
provider "aws" {
  region = "<region>"
}
```

```
data "aws_caller_identity" "current" {}

data "aws_region" "current" {}

data "aws_eks_cluster" "cluster" {
  name = var.cluster_name
}

data "aws_eks_cluster_auth" "cluster" {
  name = var.cluster_name
}

provider "kubernetes" {
  host                   = data.aws_eks_cluster.cluster.
                           endpoint
  cluster_ca_certificate = base64decode(data.aws_eks_
                           cluster.cluster.certificate_
                           authority.0.data)
  token                  = data.aws_eks_cluster_auth.
                           cluster.token
}

provider "helm" {
  kubernetes {
    host                   = data.aws_eks_cluster.cluster.
                             endpoint
    cluster_ca_certificate = base64decode(data.aws_eks_
                             cluster.cluster.certificate_
                             authority.0.data)
    token                  = data.aws_eks_cluster_auth.
                             cluster.token
  }
}
```

CHAPTER 6   THE STRUCTURE OF APPLICATION

Explanation of the Code in main.tf

- **Terraform Backend Configuration**:

    The terraform block configures the backend to store your Terraform state file in an S3 bucket. This allows you to manage and share state securely and consistently. The bucket is the name of your S3 bucket, key is the path to your state file within the bucket, and region specifies the AWS region where the bucket is located.

- **AWS Provider**:

    The provider "aws" block sets up AWS as the provider for this configuration, specifying the region where your resources will be created.

- **Data Sources**:

    - aws_caller_identity: Retrieves details about the identity making the request, such as the AWS account ID.
    - aws_region: Fetches the current region configured for AWS.
    - aws_eks_cluster: Retrieves details about your existing EKS cluster, such as its endpoint and certificate.
    - aws_eks_cluster_auth: Obtains the authentication token for your EKS cluster.

- **Kubernetes Provider**:

    The provider "kubernetes" block configures the Kubernetes provider, allowing Terraform to manage Kubernetes resources. It uses the EKS cluster's endpoint, certificate authority data (decoded from base64), and the authentication token to connect to the cluster.

- **Helm Provider**:

    Similar to the Kubernetes provider, the `provider "helm"` block is set up to manage Helm charts within your Kubernetes cluster. It also uses the EKS cluster's endpoint, certificate, and token for authentication.

The Power of EKS Add-ons

In your EKS cluster, there are a few critical components that keep the engine running smoothly—networking, DNS, and proxying. AWS has packaged these as managed add-ons, and with Terraform, we can ensure these essential pieces are always in place and up-to-date without you needing to get into the weeds.

Put the following in `addons.tf`:

```
resource "helm_release" "cni" {
  name       = "cni"
  repository = "https://aws.github.io/eks-charts"
  chart      = "aws-vpc-cni"
  namespace  = "kube-system"
  version    = "1.18.3"
}

resource "aws_eks_addon" "core-dns" {
  cluster_name = var.cluster_name
  addon_name   = "coredns"
}

resource "aws_eks_addon" "kube-proxy" {
  cluster_name = var.cluster_name
  addon_name   = "kube-proxy"
}
```

CHAPTER 6   THE STRUCTURE OF APPLICATION

What's Happening Here?

- **EKS Add-ons**: These are fully managed by AWS, meaning AWS takes care of updates, security patches, and any necessary configuration tweaks. That's less work for you and your team, letting you focus on building and scaling applications rather than managing the underpinnings.

- **VPC-CNI**: This add-on is your network plugin, the VPC-CNI. It's what handles the networking between your pods and nodes, ensuring that traffic flows smoothly within your cluster. When you deploy the VPC-CNI add-on with Terraform, you're locking in AWS's best practices for networking right out of the gate. As a bonus, each of your Kubernetes Pod gets an IP address from the VPC range. Note: You could use aws_eks_addon to introduce it to the cluster, but I found that using helm to install VPC-CNI is a bit more convenient—easier to upgrade.

- **CoreDNS**: Next up is CoreDNS. This is your cluster's DNS service. It translates service names into IP addresses, making sure that your microservices can find and talk to each other without a hitch.

- **Kube-Proxy**: Finally, there's kube-proxy. This component manages the network rules on your nodes, allowing your services to communicate both inside and outside the cluster. By deploying this as an AWS-managed add-on, you ensure that your proxying is always up-to-date, secure, and optimized for performance.

CHAPTER 6   THE STRUCTURE OF APPLICATION

Making Our Lives Easier with Helm

In the world of Kubernetes, Helm is your go-to tool for managing applications. Think of Helm as the Kubernetes equivalent of tools like apt for Ubuntu or yum for Red Hat Linux. It simplifies the process of defining, installing, and upgrading even the most complex Kubernetes applications.

How Does Helm Work?

Helm uses **charts**—pre-packaged applications or services that are easy to deploy. A Helm chart is a collection of YAML files that describe the Kubernetes resources needed to deploy an application or service. These resources might include deployments, services, ConfigMaps, secrets, and more.

Key Components of Helm

1. **Helm CLI**: The command-line interface where you run Helm commands to install, manage, and update charts in your Kubernetes cluster.

2. **Charts**: These are like blueprints for your applications. A Helm chart contains all the resource definitions necessary to run an application, tool, or service inside a Kubernetes cluster. You can think of it as a packaged application with its configuration parameters.

3. **Repositories**: Helm charts are stored in repositories. These repositories are like app stores for Kubernetes. You can pull charts from public repositories like the official Helm chart repository, or you can create your own private repositories.

4. **Releases**: When you install a chart, Helm creates a release. A release is a specific instance of a chart running in your cluster. You can have multiple releases of the same chart running in a cluster, each with its own configuration.

Why Use Helm?

- **Simplifies Deployments**: Helm makes it easy to deploy applications by abstracting complex Kubernetes configurations into simple, reusable templates.

- **Version Control**: Helm keeps track of versions of your deployments, making it easier to roll back to a previous state if something goes wrong.

- **Customization**: Helm allows you to customize your deployments using variables. You can define different values for different environments (e.g., development, staging, production) without modifying the core chart.

- **Sharing and Reuse**: Charts can be shared with the community or within your organization, promoting reuse and standardization across teams and projects.

Why We Use Helm from Terraform Instead of the Helm CLI

While Helm CLI is powerful, integrating Helm with Terraform brings several key advantages, especially when you're managing infrastructure as code:

1. **Consistency**: By using Terraform to manage Helm deployments, you keep all your infrastructure and application configurations in one place. This ensures that your Kubernetes resources and Helm releases are applied consistently and tracked together in your Terraform state.

2. **Automation**: Helm, when used from Terraform, allows for seamless automation. You can deploy and manage Kubernetes applications automatically as part of your Terraform runs, reducing the need for manual interventions and potential errors.

3. **Version Control**: Terraform provides a robust mechanism for versioning your infrastructure, including your Helm releases. This makes it easier to manage, review, and roll back changes if necessary, all within the same Terraform workflows.

4. **GitOps**: By managing Helm deployments with Terraform, you align more closely with GitOps principles. Your Kubernetes deployments become part of your codebase, where every change is tracked, reviewed, and managed via pull requests, ensuring a fully auditable and reversible deployment process.

In Summary

Helm is a powerful tool in the Kubernetes ecosystem that simplifies the deployment and management of applications. However, by integrating Helm with Terraform, you gain additional benefits like consistency,

CHAPTER 6   THE STRUCTURE OF APPLICATION

automation, version control, and alignment with GitOps practices. This combination allows you to manage your entire infrastructure, from the base layers to application deployments, in a unified, efficient, and scalable way.

In the next chapter, we are going to continue to set up foundational artifacts working on the same level using mostly Helm from Terraform. Ready when you are.

CHAPTER 7

Be a Good Kubernetes Citizen

Common Kubernetes Resources Part 2
What Is a Storage Class?

In Kubernetes, a **storage class** is a way to define the "classes" of storage available in your cluster. It abstracts the underlying storage provider (e.g., AWS EBS, Azure Disk, Google Persistent Disk) and provides a way to dynamically provision storage volumes. A storage class defines the type of storage (like SSDs or spinning disks), the performance characteristics, and other parameters that determine how the storage behaves.

When a developer or application needs storage, they don't have to worry about the specifics of how to provision it; instead, they can specify a storage class in their PersistentVolumeClaim (PVC), and Kubernetes will handle the rest, provisioning the storage according to the rules set in that storage class.

CHAPTER 7   BE A GOOD KUBERNETES CITIZEN

Why Do We Need a Custom Storage Class for EBS?

By default, when you create an EKS cluster, it comes with a predefined gp2 storage class. This default storage class uses the kubernetes.io/aws-ebs storage provisioner to create Amazon EBS volumes. However, this default storage class is somewhat limited:

1. **Flexibility**: The default gp2 storage class is set up with predefined parameters that may not meet all your requirements, such as the ability to expand volumes or set specific mount options.

2. **Provisioner Upgrade**: The kubernetes.io/aws-ebs provisioner used by the default gp2 storage class is now deprecated in favor of the newer CSI (Container Storage Interface) drivers. The AWS EBS CSI driver is more advanced, offering better features, flexibility, and support for newer EBS features.

3. **Compatibility and Future-Proofing**: AWS and Kubernetes are moving towards the use of CSI drivers for storage management because CSI is a standardized interface for container orchestrators to expose arbitrary storage systems to their containers. By adopting the CSI driver, you're future-proofing your infrastructure and aligning with best practices.

Terraform Code for Kubernetes Storage

Please put the following code into storage.tf.

1. Helm Release for AWS EBS CSI Driver

```
resource "helm_release" "ebs_csi_driver" {
  name       = "aws-ebs-csi-driver"
  namespace  = "kube-system"
  chart      = "aws-ebs-csi-driver"
  repository = "https://kubernetes-sigs.github.io/aws-ebs-
               csi-driver/"
  version    = "2.25.0"
}
```

- **Helm Release**: This block installs the AWS EBS CSI driver in your EKS cluster using Helm, which is the package manager for Kubernetes.

- **Name and Namespace**: The driver is installed in the kube-system namespace, which is where critical system-level components reside.

- **Chart and Repository**: You're pulling the AWS EBS CSI driver from its official Helm chart repository and specifying version 2.25.0.

This installation is necessary because the default gp2 storage class uses the older in-tree provisioner (kubernetes.io/aws-ebs), whereas the AWS EBS CSI driver allows Kubernetes to interface with AWS EBS volumes in a more flexible and feature-rich way.

2. Custom Storage Class

```
resource "kubernetes_storage_class" "ebs" {
  metadata {
    name = var.storage_class
    annotations = {
```

```
    "storageclass.kubernetes.io/is-default-class" = "true"
  }
}
storage_provisioner = "kubernetes.io/aws-ebs"
parameters = {
  type = "gp2"
}
reclaim_policy = "Delete"
allow_volume_expansion = true
mount_options = ["debug"]
volume_binding_mode = "Immediate"
depends_on = [
  helm_release.ebs_csi_driver
]
}
```

- **Name**: The storage class is named ebs, which makes it easy to refer to in PersistentVolumeClaims.

- **Provisioner**: Initially, you might see kubernetes.io/aws-ebs, but after installing the CSI driver, you'll likely want to use ebs.csi.aws.com as the provisioner, depending on your configuration. The use of kubernetes.io/aws-ebs here is likely a placeholder or for compatibility, but using the CSI driver is recommended.

- **Parameters**: Specifies gp2 as the type of EBS volume.

CHAPTER 7   BE A GOOD KUBERNETES CITIZEN

- **Reclaim Policy**: When a PersistentVolume (PV) created by this storage class is deleted, its backing storage in AWS will also be deleted (`Delete`).

- **Allow Volume Expansion**: This allows the volume to be resized dynamically if your application needs more storage.

- **Mount Options**: Adding options like `debug` can be useful for troubleshooting.

- **Volume Binding Mode**: `Immediate` binding mode means that the PV will be created and bound to a PVC as soon as the claim is created, rather than waiting until a pod using the PVC is scheduled.

- **Depends On**: The `depends_on` block ensures that the AWS EBS CSI driver is installed before this storage class is created, as the driver is necessary for the class to function correctly.

Why Not Use the Default gp2 Storage Class?

- **Limited Features**: The default storage class doesn't allow you to take advantage of newer features provided by the AWS EBS CSI driver, such as dynamic volume resizing, improved performance, and advanced configurations.

- **Deprecated Provisioner**: The `kubernetes.io/aws-ebs` provisioner used in the default `gp2` storage class is being phased out. The CSI driver (`ebs.csi.aws.com`) is the future of storage management in Kubernetes, providing better compatibility with evolving Kubernetes features and AWS services.

- **Customization**: With a custom storage class, you can tailor the storage behavior to your specific application needs, like enabling volume expansion, setting mount options, or controlling the reclaim policy.

Conclusion on Storage

By using Terraform to set up a custom storage class with the AWS EBS CSI Driver, you're ensuring that your EKS cluster is using the latest, most feature-rich storage solution available. This approach not only future-proofs your infrastructure but also provides you with the flexibility to manage your storage resources more effectively, avoiding the limitations of the default gp2 storage class.

Karpenter Application

We have some preparations on previous steps for what happens next—setup of Karpenter application itself. Please put the following in karpenter.tf:

```
resource "kubernetes_namespace" "karpenter" {
  metadata {
    name = "karpenter"
  }
}
```

- **Purpose**: This block creates a new Kubernetes namespace named karpenter within your Kubernetes cluster.
- **Namespace**: A namespace in Kubernetes is used to organize and separate resources. By creating a specific namespace for Karpenter, you isolate its resources from other components in the cluster, making management easier and more secure.

CHAPTER 7   BE A GOOD KUBERNETES CITIZEN

```
resource "helm_release" "karpenter" {
  name       = "karpenter"
  repository = "oci://public.ecr.aws/karpenter"
  chart      = "karpenter"
  version    = var.karpenter_version
  namespace  = kubernetes_namespace.karpenter.
               metadata[0].name

  set {
    name  = "settings.clusterName"
    value = var.cluster_name
  }

  set {
    name  = "settings.interruptionQueue"
    value = var.cluster_name
  }

  set {
    name  = "controller.resources.requests.cpu"
    value = "1"
  }

  set {
    name  = "controller.resources.requests.memory"
    value = "1Gi"
  }

  set {
    name  = "controller.resources.limits.cpu"
    value = "1"
  }
```

65

```
    set {
      name  = "controller.resources.limits.memory"
      value = "1Gi"
    }
    set {
      name  = "nodeSelector.${var.karpenter_master_node_
              label_name}"
      value = var.karpenter_master_node_label_value
    }
    wait_for_jobs = true
  }
```

- **Helm Release**: This block installs Karpenter into your Kubernetes cluster using Helm, a package manager for Kubernetes.

- **Name and Repository**:
  - **name**: The release is named `karpenter`.
  - **repository**: Specifies the OCI-compliant repository where the Karpenter Helm chart is stored, in this case, hosted on AWS's public ECR (Elastic Container Registry).

- **Chart and Version**:
  - **chart**: The specific chart being used to deploy Karpenter.
  - **version**: The version of the Karpenter chart being deployed, which is controlled by the `var.karpenter_version` variable.

- **Namespace**:
  - The Helm release is deployed into the `karpenter` namespace created earlier. This is referenced dynamically using `kubernetes_namespace.karpenter.metadata[0].name`.

CHAPTER 7   BE A GOOD KUBERNETES CITIZEN

- **Set Commands**:
  - These set blocks customize the deployment by overriding default values in the Helm chart with specific values. Here's what each does:
    - **settings.clusterName**: Sets the name of the EKS cluster to which Karpenter will be connected, using `var.cluster_name`.
    - **settings.interruptionQueue**: Configures Karpenter to use a specific interruption queue, also set to `var.cluster_name`.
    - **controller.resources.requests.cpu and controller.resources.requests.memory**: Specifies the minimum CPU and memory resources requested by the Karpenter controller.
    - **controller.resources.limits.cpu and controller.resources.limits.memory**: Sets the maximum CPU and memory resources that the Karpenter controller can use.
    - **nodeSelector**: Configures Karpenter to run on nodes with a specific label, using `var.karpenter_master_node_label_name` and `var.karpenter_master_node_label_value`.

- **wait_for_jobs**:
  - This setting ensures that Terraform waits for any jobs created by the Helm chart to complete before continuing. This is important for ensuring that Karpenter is fully initialized and ready before Terraform moves on to the next tasks.

Summary on Karpenter

This Terraform configuration sets up Karpenter, a Kubernetes cluster autoscaler, within your Kubernetes cluster. It creates a dedicated namespace for Karpenter and deploys it using Helm, with several custom configurations to tailor the deployment to your specific cluster environment. The use of `nodeSelector`, resource requests, and limits ensures that Karpenter is efficiently managed and operates within the boundaries you define. The deployment is handled in a declarative, repeatable manner, aligning with best practices for infrastructure as code.

Why Use an Ingress Object?

In a Kubernetes cluster, while you can expose services directly using a `service` of type `LoadBalancer` or `NodePort`, this approach is generally not recommended for application developers. Instead, you should use an **Ingress** object, which provides a more efficient and scalable way to manage external access to your services. The Ingress object allows you to define rules for routing traffic to different services within your cluster based on the URL path or hostname, enabling more sophisticated and centralized traffic management.

What Is an Ingress Controller?

To make the Ingress object functional, you need an **Ingress Controller**. The Ingress Controller is responsible for interpreting the rules defined in the Ingress object and managing the underlying networking to route traffic accordingly. It acts as a gateway, handling tasks like SSL termination, load balancing, and path-based routing, effectively bridging external traffic to your internal Kubernetes services.

CHAPTER 7   BE A GOOD KUBERNETES CITIZEN

Why Use the ALB Ingress Controller?

We're going to use the **AWS ALB Ingress Controller** because it integrates seamlessly with AWS's Application Load Balancer (ALB). This controller allows you to leverage ALB's advanced features, such as path-based routing, SSL termination, and Web Application Firewall (WAF) support, directly within your Kubernetes cluster. Using the ALB Ingress Controller ensures that your applications are managed securely and efficiently while taking full advantage of AWS's powerful networking capabilities.

Let's Install ALB Ingress Controller

Please create ingress.tf and put the following content:

```
resource "helm_release" "alb_ingress" {
  name       = "alb"
  repository = "https://aws.github.io/eks-charts"
  chart      = "aws-load-balancer-controller"
  version    = "1.6.2"
  namespace  = "kube-system"
  set {
    name  = "clusterName"
    value = var.cluster_name
  }
}
```

External DNS

You'll see shortly how we are going to use it, right now we just need to have it. Please create external_dns.tf and put the following content into

```
locals {
    sa_name = "external-dns" // k8s service account named
    sa_ns   = "kube-system"  // namespace where the service
                                account is
}
resource "aws_iam_role" "external_dns_role" {
  name = "external-dns-role-${var.cluster_name}"

  assume_role_policy = data.aws_iam_policy_document.external_
                      dns_assume_role_policy.json
}

data "aws_iam_openid_connect_provider" "eks_openid_connect_
provider" {
  url = data.aws_eks_cluster.cluster.identity.0.oidc.0.issuer
}

data "aws_iam_policy_document" "external_dns_assume_role_
policy" {
  statement {
    actions = ["sts:AssumeRoleWithWebIdentity"]
    effect  = "Allow"
    principals {
      type        = "Federated"
      identifiers = [data.aws_iam_openid_connect_provider.eks_
                    openid_connect_provider.arn]
    }
```

```
    condition {
      test     = "StringEquals"
      variable = "${data.aws_iam_openid_connect_provider.eks_
                  openid_connect_provider.url}:sub"
      values   = ["system:serviceaccount:${local.sa_
                  ns}:${local.sa_name}"]
    }
  }
}
resource "aws_iam_policy" "external_dns_policy" {
  name = "ExternalDNSPolicy-${var.cluster_name}"

  policy = jsonencode({
    Version = "2012-10-17"
    Statement = [
      {
        Effect = "Allow"
        Action = [
          "route53:ChangeResourceRecordSets",
          "route53:ListHostedZones",
          "route53:ListResourceRecordSets"
        ]
        Resource = "*"
      },
      {
        Effect = "Allow"
        Action = "route53:GetChange"
        Resource = "arn:aws:route53:::change/*"
      }
    ]
  })
}
```

```
resource "aws_iam_role_policy_attachment" "external_dns_role_attachment" {
  role       = aws_iam_role.external_dns_role.name
  policy_arn = aws_iam_policy.external_dns_policy.arn
}

resource "kubernetes_service_account" "external_dns_sa" {
  metadata {
    name      = "external-dns"
    namespace = "kube-system"
    annotations = {
      "eks.amazonaws.com/role-arn" = aws_iam_role.external_dns_role.arn
    }
  }
}

resource "helm_release" "external_dns" {
  name       = "external-dns"
  namespace  = "kube-system"
  repository = "https://kubernetes-sigs.github.io/external-dns/"
  chart      = "external-dns"

  set {
    name  = "provider"
    value = "aws"
  }

  set {
    name  = "sources[0]"
    value = "ingress"
  }
```

```
  set {
    name  = "serviceAccount.create"
    value = "false"
  }

  set {
    name  = "aws.zoneType"
    value = "public"
  }

  set {
    name  = "aws.region"
    value = data.aws_region.current.name
  }

  set {
    name  = "txtOwnerId"
    value = "external-dns"
  }

  set {
    name  = "policy"
    value = "sync"
  }

  set {
    name  = "serviceAccount.name"
    value = kubernetes_service_account.external_dns_
sa.metadata[0].name
  }
}
```

I know, it is a lot of magic here, just copy paste it for now. We will get explanations later.

CHAPTER 7   BE A GOOD KUBERNETES CITIZEN

It's Time to Apply

Run `terraform init` and `terraform apply` while you are in 01-k8s directory. Your Kubernetes cluster is almost there. Only one step left. Turn the page.

CHAPTER 8

The Intricacies of Helm

The Hard Parts

This chapter is where things get more complex, but don't worry—we'll start with the basics and build from there.

Kubectl: Your Key to the Kubernetes Cluster

kubectl is the command-line tool that allows you to interact with your Kubernetes cluster. It's your gateway to managing and troubleshooting everything within the cluster. With kubectl, you can deploy applications, inspect and manage cluster resources, and view logs—all from your terminal.

Installation: You can install kubectl by following the official documentation for your operating system. Once installed, kubectl becomes your primary tool for communicating with Kubernetes.

CHAPTER 8   THE INTRICACIES OF HELM

Kubectl Configuration

Your `kubectl` configuration is typically stored in `~/.kube/config`. This file contains the details needed to connect to your Kubernetes clusters, such as cluster endpoints and authentication tokens.

To set up `kubectl` for your EKS cluster, run the following command:

`aws eks update-kubeconfig --name <your-cluster-name>`

Then, verify your access to the cluster by running

`kubectl get nodes`

If you see a list of nodes, you're successfully connected.

Basic Kubectl Commands

Here are some of the most commonly used `kubectl` commands:

- `kubectl get pods`: List all pods in the current namespace.
- `kubectl get services`: List all services.
- `kubectl describe pod <pod-name>`: Display detailed information about a specific pod.
- `kubectl logs <pod-name>`: View logs from a specific pod.
- `kubectl apply -f <file.yaml>`: Apply a configuration from a file.
- `kubectl delete <resource-type> <resource-name>`: Delete a resource.
- `kubectl exec -it <pod-name> -- /bin/bash`: Access a shell within a running pod.

CHAPTER 8   THE INTRICACIES OF HELM

Kubectl api-resources

Running the command kubectl api-resources will list all the resources available in your Kubernetes cluster, including the API versions and resource names. This command is incredibly useful for understanding what resources are available to you and how they can be managed within the cluster.

Introducing CRDs

A **Custom Resource Definition (CRD)** allows you to extend Kubernetes by defining your own resource types. This is crucial for integrating custom functionality or third-party tools into your cluster. When you create a CRD, you're essentially teaching Kubernetes how to handle a new type of resource.

CRDs are commonly used in advanced Kubernetes setups, where you might need to manage resources that aren't natively supported by Kubernetes. These custom resources behave just like the built-in ones, and you can manage them using the same kubectl commands.

Terraform's Inconvenience with CRDs

Currently, the Terraform Kubernetes provider struggles with managing CRDs effectively, especially when it tries to get the status of a resource before the corresponding CRD has been created. This leads to errors in scenarios like the following:

```
resource "kubernetes_manifest" "nodepool" {
  manifest = {
    apiVersion = "karpenter.sh/v1beta1"
    kind       = "NodePool"
    metadata = {
```

```
      name = "default"
    }
  }
}
```

If the "karpenter.sh/v1beta1" CRD hasn't been created yet, this code will fail. This is why we have to create CRDs separately and ensure they are applied before any resources that depend on them. To work around this limitation, I've structured the Terraform code into separate folders, each handling different stages of the setup. This extra level of organization is necessary due to the current limitations of the Terraform Kubernetes provider.

Creating the 02-karpenter Folder

Your directory structure will look like this:

```
00-vpceks
01-k8s
02-karpenter
```

This structure ensures that CRDs are handled correctly and in the right order, allowing your Terraform setup to proceed smoothly without running into dependency issues.

The Secret of aws-auth ConfigMap

The `aws-auth` ConfigMap is the critical bridge between AWS IAM and Kubernetes RBAC. It allows you to map IAM roles and users to Kubernetes roles, enabling fine-grained access control within your EKS cluster. Essentially, this ConfigMap is how AWS IAM identities gain permissions to interact with Kubernetes resources.

Understanding the Content of aws-auth

The aws-auth ConfigMap contains mappings that link IAM entities (like users and roles) to Kubernetes RBAC roles. Here's an example of what the content might look like:

```
apiVersion: v1
kind: ConfigMap
metadata:
  name: aws-auth
  namespace: kube-system
data:
  mapRoles: |
    - rolearn: arn:aws:iam::123456789012:role/KubernetesAdmin
      username: admin
      groups:
        - system:masters
  mapUsers: |
    - userarn: arn:aws:iam::123456789012:user/JohnDoe
      username: johndoe
      groups:
        - system:masters
```

- **mapRoles**: Maps IAM roles to Kubernetes users and groups. For example, the KubernetesAdmin IAM role is mapped to the admin user, who is part of the system:masters group in Kubernetes, granting admin-level access.

- **mapUsers**: Maps individual IAM users to Kubernetes users and groups. For instance, the JohnDoe IAM user is mapped to the johndoe user in Kubernetes, with the same system:masters group access.

CHAPTER 8   THE INTRICACIES OF HELM

How to Read and Edit the aws-auth ConfigMap

You can view the aws-auth ConfigMap using the following kubectl command:

```
kubectl get configmap -n kube-system aws-auth -o yaml
```

You can edit the aws-auth ConfigMap using the following command:

```
kubectl edit configmap -n kube-system aws-auth
```

Changes to the aws-auth ConfigMap take effect immediately, so it's a powerful tool for controlling access to your Kubernetes cluster, directly linking your AWS IAM policies with Kubernetes' RBAC system. You could edit this way but you should not, we are going to use Terraform for that task.

Time to Create Main.tf

Put the following in 02-karpenter/main.tf.

Backend Configuration

- **Purpose**: The terraform block configures the backend to store the Terraform state file in an S3 bucket. This allows you to manage and share state securely across your team. The bucket is the name of your S3 bucket, key specifies the path to your state file within the bucket, and region indicates the AWS region where the bucket is located.

```
terraform {
  backend "s3" {
    bucket = "<state-bucket-name>"
    key    = "02-karpenter.tfstate"
    region = "<region>"
  }
}
provider "aws" {
  region = "<region>"
}
```

Cluster Information and Kubernetes Provider

- **Purpose**:

  - **Variable**: The cluster_name variable holds the name of your EKS cluster.

  - **Data Sources:** These data sources fetch details about your EKS cluster, such as its endpoint and authentication token.

  - **Kubernetes Provider**: The provider "kubernetes" block configures Terraform to interact with your Kubernetes cluster using the endpoint, certificate, and token retrieved from the data sources.

    ```
    variable "cluster_name" {
      default = "<your-cluster-name>"
    }
    ```

```
data "aws_eks_cluster" "cluster" {
  name = var.cluster_name
}
data "aws_eks_cluster_auth" "cluster" {
  name = var.cluster_name
}
provider "kubernetes" {
  host                   = data.aws_eks_cluster.
                           cluster.endpoint
  cluster_ca_certificate = base64decode(data.
                           aws_eks_cluster.
                           cluster.certificate_
                           authority.0.data)
  token                  = data.aws_eks_cluster_
                           auth.cluster.token
}
```

IAM Role for Karpenter Nodes

- **Purpose**: This resource creates an IAM role for Karpenter nodes, allowing them to assume specific policies needed for operation.
  - **Inline Policy**: Grants permissions related to AWS Shield, WAFv2, and WAF Regional services.
  - **Assume Role Policy**: Allows EC2 instances to assume this role, which is necessary for Karpenter-managed nodes to operate correctly.

```
resource "aws_iam_role" "karpenter_node_role" {
  name = "KarpenterNodeRole${var.cluster_name}"
  inline_policy {
    name = "KarpenterNodePolicy"
    policy = jsonencode({
      Version = "2012-10-17",
      Statement = [
        {
          Effect   = "Allow",
          Action   = ["shield:Get*"],
          Resource = ["*"]
        },
        {
          Effect   = "Allow",
          Action   = ["wafv2:Get*"],
          Resource = ["*"]
        },
        {
          Effect   = "Allow",
          Action   = ["waf-regional:Get*"],
          Resource = ["*"]
        }
      ]
    })
  }
  assume_role_policy = jsonencode({
    Version = "2012-10-17",
    Statement = [
      {
        Effect = "Allow",
        Principal = {
          Service = "ec2.amazonaws.com"
```

```
        },
        Action = "sts:AssumeRole"
      }
    ]
  })
}
```

Attaching Policies to the IAM Role

- **Purpose**: These resources attach essential AWS-managed policies to the Karpenter node IAM role, enabling it to

    - Interact with EKS (`AmazonEKSWorkerNodePolicy`).

    - Manage network interfaces (`AmazonEKS_CNI_Policy`).

    - Access ECR for pulling container images (`AmazonEC2ContainerRegistryReadOnly`).

    - Interact with SSM (`AmazonSSMManagedInstanceCore`).

    - Manage load balancers (`ElasticLoadBalancingFullAccess`).

    ```
    resource "aws_iam_role_policy_attachment" "eks_worker_node_policy" {
      role       = aws_iam_role.karpenter_node_role.name
      policy_arn = "arn:aws:iam::aws:policy/
                   AmazonEKSWorkerNodePolicy"
    }

    resource "aws_iam_role_policy_attachment" "eks_cni_policy" {
      role       = aws_iam_role.karpenter_node_role.name
    ```

```
    policy_arn = "arn:aws:iam::aws:policy/AmazonEKS_
                  CNI_Policy"
}

resource "aws_iam_role_policy_attachment" "ecr_
read_only" {
  role       = aws_iam_role.karpenter_node_role.name
  policy_arn = "arn:aws:iam::aws:policy/AmazonEC2
                ContainerRegistryReadOnly"
}

resource "aws_iam_role_policy_attachment" "ssm_
managed_instance_core" {
  role       = aws_iam_role.karpenter_node_role.name
  policy_arn = "arn:aws:iam::aws:policy/
                AmazonSSMManagedInstanceCore"
}

resource "aws_iam_role_policy_attachment" "lb" {
  role       = aws_iam_role.karpenter_node_role.name
  policy_arn = "arn:aws:iam::aws:policy/ElasticLoad
                BalancingFullAccess"
}
```

Creating Kubernetes Resources for Karpenter

- **Purpose**: This resource creates a NodePool in Kubernetes using Karpenter, which manages the lifecycle of the nodes.
  - **Requirements**: Specifies the architecture (amd64), operating system (linux), and instance characteristics (e.g., instance category, generation) that nodes in this pool must meet.

CHAPTER 8   THE INTRICACIES OF HELM

- **NodeClassRef**: Refers to the EC2NodeClass resource that defines the specifics of the EC2 instances to be used.
- **Limits and Disruption**: Sets CPU limits and defines a policy for consolidating underutilized nodes, along with an expiration time for nodes.

```
resource "kubernetes_manifest" "nodepool" {
  manifest = {
    apiVersion = "karpenter.sh/v1beta1"
    kind       = "NodePool"
    metadata = {
      name = "default"
    }
    spec = {
      template = {
        spec = {
          requirements = [
            {
              key      = "kubernetes.io/arch"
              operator = "In"
              values   = ["amd64"]
            },
            {
              key      = "kubernetes.io/os"
              operator = "In"
              values   = ["linux"]
            },
            {
              key      = "karpenter.sh/capacity-type"
              operator = "In"
              values   = ["spot"]
            },
```

```
          {
            key      = "karpenter.k8s.aws/
                        instance-category"
            operator = "In"
            values   = ["c", "m", "r"]
          },
          {
            key      = "karpenter.k8s.aws/instance-
                        generation"
            operator = "Gt"
            values   = ["2"]
          }
        ]
        nodeClassRef = {
          name = "default"
        }
      }
    }
    limits = {
      cpu = "1k"
    }
    disruption = {
      consolidationPolicy = "WhenUnderutilized"
      expireAfter         = "720h" # 30 *
                                    24h = 720h
    }
   }
  }
 }
}
```

CHAPTER 8   THE INTRICACIES OF HELM

Creating EC2NodeClass Resource for Karpenter

- **Purpose**: This resource defines an EC2NodeClass, specifying the details of the EC2 instances Karpenter will use for the NodePool.
  - **AMI Family**: Specifies the operating system to use (Amazon Linux 2).
  - **Role**: Associates the previously created IAM role with these instances.
  - **Subnet and Security Group Selectors**: Defines which subnets and security groups the instances will be part of, using tags for discovery.

```
resource "kubernetes_manifest" "ec2nodeclass" {
  manifest = {
    apiVersion = "karpenter.k8s.aws/v1beta1"
    kind       = "EC2NodeClass"
    metadata = {
      name = "default"
    }
    spec = {
      amiFamily = "AL2" # Amazon Linux 2
      role      = aws_iam_role.karpenter_node_
                  role.name
      subnetSelectorTerms = [
        {
          tags = {
            "karpenter.sh/discovery" = var.cluster_name
          }
        }
```

```
      ]
      securityGroupSelectorTerms = [
        {
          tags = {
            "karpenter.sh/discovery" = var.cluster_name
          }
        }
      ]
    }
  }
}
```

Updating the aws-auth ConfigMap

- **Purpose**: This code updates the `aws-auth` ConfigMap to map the newly created IAM role to the Kubernetes `system:nodes` group, allowing Karpenter-managed EC2 instances to join the Kubernetes cluster and operate as nodes.

    - **mapRoles**: Appends the Karpenter node role to the existing roles in the ConfigMap, ensuring that these nodes have the necessary permissions to function within the cluster.

      ```
      data "kubernetes_config_map" "aws_auth" {
        metadata {
          name      = "aws-auth"
          namespace = "kube-system"
        }
      }

      resource "kubernetes_config_map_v1_data" "aws_auth" {
        metadata {
      ```

```
      name      = "aws-auth"
      namespace = "kube-system"
    }
    force = true
    lifecycle {
      prevent_destroy = true
    }
    data = {
      "mapRoles" = yamlencode(
        [
          yamldecode(data.kubernetes_config_map.aws_
          auth.data["mapRoles"])[0],
          {
            "groups" = [
              "system:bootstrappers",
              "system:nodes",
            ]
            "rolearn"  = "${aws_iam_role.karpenter_
                          node_role.arn}"
            "username" = "system:node:{{EC2Private
                          DNSName}}"
          },
        ]
      )
    }
  }
```

Put in Action

Run `terraform init` followed by `terraform apply` while you navigated to 02-karpenter folder.

CHAPTER 9

Choosing the Right CI/CD Platform

Setting GitLab in Action

Why GitLab?

GitLab is more than just a Git repository manager—it's a comprehensive DevOps platform that integrates source control, CI/CD pipelines, and even project management tools all in one place. Developers appreciate GitLab for its flexibility and powerful features that go beyond what GitHub or Bitbucket typically offer. One of GitLab's standout features is its built-in CI/CD capabilities, which allow developers to automate testing, deployment, and more directly from within the platform.

Moreover, GitLab's commitment to open source is a big draw for the developer community. Unlike GitHub or Bitbucket, which have more restricted free tiers, GitLab's open-core model ensures that a significant portion of its features is available to everyone, fostering a vibrant community of contributors. This community-driven development not only keeps GitLab innovative but also makes it a tool that grows and evolves with the needs of its users.

GitLab Components

When you set up a GitLab cluster, you're not just deploying a single application but rather a collection of services that work together to provide a seamless developer experience. Key components of a GitLab cluster include

- **GitLab Rails (Application Server)**: The core component that handles the web interface, API requests, and background jobs.

- **Gitaly**: Manages Git repositories and serves Git data to GitLab Rails.

- **PostgreSQL**: The database backend that stores all GitLab data, including user accounts, projects, and CI/CD pipelines.

- **Redis**: Used for caching, background jobs, and session storage to improve performance and scalability.

- **Sidekiq**: Processes background jobs, handling everything from sending emails to running CI/CD jobs.

Each of these components is crucial for the overall functionality of the GitLab platform, ensuring that developers can manage their projects, collaborate on code, and automate deployments all in one place.

Use Helm to Install GitLab

As with our usual approach, we're going to install the GitLab cluster using Helm and Terraform. This method provides the flexibility of Helm's package management combined with Terraform's infrastructure as code capabilities. This time, we'll be using a values file to customize the Helm installation to fit our specific needs.

CHAPTER 9 CHOOSING THE RIGHT CI/CD PLATFORM

This file allows us to fine-tune the configuration of the GitLab components, ensuring that our deployment is optimized for performance, scalability, and security. By leveraging Helm with Terraform, we ensure that our GitLab installation is not only robust but also easily reproducible and manageable within our existing infrastructure.

Let's Start

Please create a folder 03-gitlab, so it will be

```
00-vpceks
01-k8s
02-karpenter
03-gitlab
```

Put main.tf

It is our usual content for main.tf.

```
terraform {
  backend "s3" {
    bucket = "<bucket-for-state>"
    key    = "03-gitlab.tfstate"
    region = "<region>"
  }
}
provider "aws" {
  region = "<region>"
}
data "aws_eks_cluster" "cluster" {
  name = var.cluster_name
}
```

```
data "aws_eks_cluster_auth" "cluster" {
  name = var.cluster_name
}
provider "kubernetes" {
  host                   = data.aws_eks_cluster.cluster.
                           endpoint
  cluster_ca_certificate = base64decode(data.aws_eks_
                           cluster.cluster.certificate_
                           authority.0.data)
  token                  = data.aws_eks_cluster_auth.
                           cluster.token
}
provider "helm" {
  kubernetes {
    host                   = data.aws_eks_cluster.cluster.
                             endpoint
    cluster_ca_certificate = base64decode(data.aws_eks_
                             cluster.cluster.certificate_
                             authority.0.data)
    token                  = data.aws_eks_cluster_auth.
                             cluster.token
  }
}
```

variables.tf

I believe the following is self-explanatory:

```
variable "storage_class_name" {
    default = "ebs"
}
```

```
variable "cluster_name" {
    default = "<your-cluster-name>"
}
variable "domain" {
    default = "<your-domain>"    // use could use the same value
                                 as in 00-vpceks/variables.tf
}
```

outputs.tf

```
output "gitlab_pwd" {
  sensitive = true
  value = data.kubernetes_secret.gitlab_pwd.data.password
}
```

After terraform apply you can get the password by running `terraform output gitlab_pwd` and use it for login with the user name root.

Terraform Template Files

Terraform template files (*.tpl) are the secret sauce for injecting dynamic, context-specific configurations into your infrastructure deployments. When managing complex setups like a GitLab installation with Helm, template files become indispensable. They allow you to generate configuration files—like the values.yaml for Helm—tailored precisely to your environment. Instead of hardcoding values that might change across environments, you can use variables and expressions within the template

CHAPTER 9   CHOOSING THE RIGHT CI/CD PLATFORM

gitlab-values.tpl

```
certmanager:
  install: false

certmanager-issuer:
  email: "dummy@foobar.com"
  create: false

nginx-ingress:
  enabled: false

global:
  hosts:
    domain: ${domain}
  edition: ce
  ingress:
    enabled: true
    class: alb
    annotations:
      kubernetes.io/ingress.class: "alb"
      alb.ingress.kubernetes.io/scheme: "internet-facing"
      alb.ingress.kubernetes.io/target-type: "ip"
      alb.ingress.kubernetes.io/certificate-arn:
      "${certificate_arn}"
      alb.ingress.kubernetes.io/listen-ports: '[{"HTTP": 80},
      {"HTTPS": 443}]'
      alb.ingress.kubernetes.io/group.name: "${group_name}"
      alb.ingress.kubernetes.io/ssl-redirect: "443"
      alb.ingress.kubernetes.io/healthcheck-path: "/login"

gitlab-runner:
  runners:
    cache:
```

```
      cacheShared: true
      storageClass: ${storage_class}
redis:
  master:
    persistence:
      storageClass: ${storage_class}
  slave:
    persistence:
      storageClass: ${storage_class}
minio:
  persistence:
    storageClass: ${storage_class}
```

Understanding the GitLab Values File

This file is a configuration blueprint that customizes the deployment of GitLab using Helm. Instead of using the default configurations, this file tailors the setup to better fit your specific infrastructure, particularly by leveraging the AWS ALB Ingress Controller instead of the default NGINX Ingress Controller and Cert Manager.

Key Sections and Configuration

- **certmanager**:
  - `install: false`: This setting disables the installation of Cert Manager. By default, GitLab might attempt to use Cert Manager for managing SSL certificates, but in this setup, we are not relying on it since we're handling SSL with AWS's ALB.

CHAPTER 9   CHOOSING THE RIGHT CI/CD PLATFORM

- **nginx-ingress**:
  - `enabled: false`: This explicitly disables the NGINX Ingress Controller, which is often used by default in Kubernetes environments. Instead, we're utilizing the AWS ALB Ingress Controller that we configured in previous chapters.

- **global**:
  - **hosts**:
    - `domain: ${domain}`: Specifies the domain for GitLab, dynamically set through a variable.
  - **edition**:
    - ce: Deploys the Community Edition (CE) of GitLab.
  - **ingress**:
    - `enabled: true`: Enables the Ingress configuration.
    - `class: alb`: Specifies that the Ingress should use the ALB (Application Load Balancer) class.
    - **annotations**: Annotations are key-value pairs that provide metadata to Kubernetes resources, often instructing how they should behave. In this case, they configure specific behaviors for the ALB:
      - `kubernetes.io/ingress.class: "alb"`: Directs Kubernetes to use the ALB Ingress class.
      - `alb.ingress.kubernetes.io/scheme: "internet-facing"`: Configures the ALB to be accessible from the internet.

CHAPTER 9   CHOOSING THE RIGHT CI/CD PLATFORM

- `alb.ingress.kubernetes.io/target-type: "ip"`: Specifies that the ALB will route traffic directly to IP addresses of the pods.

- `alb.ingress.kubernetes.io/certificate-arn: "${certificate_arn}"`: Associates an SSL certificate with the ALB for HTTPS traffic, dynamically set through a variable.

- `alb.ingress.kubernetes.io/listen-ports: '[{"HTTP": 80}, {"HTTPS": 443}]'`: Configures the ALB to listen on both HTTP and HTTPS ports.

- `alb.ingress.kubernetes.io/group.name: "${group_name}"`: Assigns the ALB to a specific group, useful for managing multiple Ingress resources.

- `alb.ingress.kubernetes.io/ssl-redirect: "443"`: Forces HTTP traffic to be redirected to HTTPS.

- `alb.ingress.kubernetes.io/healthcheck-path: "/login"`: Sets the health check path to ensure the ALB can monitor the health of the GitLab service.

- **gitaly**:
  - **persistence:**
    - `storageClass: ${storage_class}`: Configures persistent storage for Gitaly using the specified storage class, which is dynamically set.

- **postgresql**:
  - **persistence:**
    - `storageClass: ${storage_class}`: Similar to Gitaly, this sets up persistent storage for PostgreSQL, GitLab's database, using the specified storage class.
- **gitlab-runner**:
  - **runners:**
    - **cache**:
      - `cacheShared: true`: Enables a shared cache for GitLab runners.
      - `storageClass: ${storage_class}`: Defines the storage class for caching, again using the variable to dynamically set the storage class.
- **redis**:
  - **master**:
    - **persistence:**
      - `storageClass: ${storage_class}`: Configures persistent storage for the Redis master.
  - **slave:**
    - **persistence**:
      - `storageClass: ${storage_class}`: Configures persistent storage for Redis slaves.

CHAPTER 9 CHOOSING THE RIGHT CI/CD PLATFORM

- **minio**:
  - **persistence**:
    - `storageClass: ${storage_class}`: Sets up persistent storage for MinIO, the object storage service used by GitLab.

gitlab.tf
Creating a Kubernetes Namespace

- **Purpose**: This block creates a Kubernetes namespace named `gitlab`. Namespaces in Kubernetes are used to isolate resources within the same cluster, providing a way to manage and organize resources like services, pods, and secrets. By creating a dedicated namespace for GitLab, we ensure that all related resources are grouped together and isolated from other applications within the cluster.

```
resource "kubernetes_namespace" "gitlab" {
  metadata {
    name = "gitlab"
  }
}
```

Fetching the SSL Certificate

- **Purpose**: This data block fetches the most recent AWS ACM (Amazon Certificate Manager) SSL certificate that is issued for the domain specified in `var.domain`. This certificate will be used to secure the GitLab instance with HTTPS. The `statuses = ["ISSUED"]` ensures that only valid, active certificates are considered. We did issue this certificate in 00-vpceks.

101

```
data "aws_acm_certificate" "gitlab" {
  domain      = var.domain
  statuses    = ["ISSUED"]
  most_recent = true
}
```

Deploying GitLab with Helm

- **Purpose**: This block deploys GitLab using Helm within the Kubernetes cluster.
  - **name**: The release is named `gitlab`.
  - **repository**: Specifies the official GitLab Helm chart repository.
  - **chart**: The specific chart to deploy is named `gitlab`.
  - **namespace**: The GitLab release is installed into the `gitlab` namespace created earlier.
  - **wait**: Ensures Terraform waits until the deployment is fully rolled out before continuing with the next steps.
  - **values**: The Helm chart is customized using a `values.yaml` file generated by the `templatefile` function. This function takes a template file (`gitlab-values.tpl`) and fills in placeholders with the provided values:
    - `domain`: Sets the domain for GitLab to `luxoft.academy`.
    - `certificate_arn`: Injects the ARN of the SSL certificate retrieved from AWS ACM.
    - `group_name`: Sets the group name for the ALB Ingress Controller.

- storage_class: Specifies the storage class to be used for persistent storage, passed in as a variable (var.storage_class_name).

```
resource "helm_release" "gitlab" {
  name       = "gitlab"
  repository = "https://charts.gitlab.io/"
  chart      = "gitlab"
  namespace  = kubernetes_namespace.gitlab.
               metadata[0].name

  wait   = false
  values = [templatefile("${path.module}/gitlab-
           values.tpl",
    {
      domain          = var.domain,
      certificate_arn = data.aws_acm_certificate.
                        gitlab.arn,
      group_name      = "gitlab"
      storage_class   = var.storage_class_name
    }
  )]
}
```

Fixing DNS

There, I should be writing Terraform code aiming to fix DNS records for our installation, but I won't. ExternalDns takes care about everything.

CHAPTER 9   CHOOSING THE RIGHT CI/CD PLATFORM

Retrieving the GitLab Root Password

- **Purpose**: This block retrieves the initial root password for the GitLab instance, which is stored as a Kubernetes secret. The secret is automatically created during the Helm deployment of GitLab.
  - **metadata**: Specifies the name and namespace of the secret to fetch (gitlab-gitlab-initial-root-password within the gitlab namespace).
  - **depends_on**: Ensures that the secret retrieval only occurs after the GitLab Helm release has been successfully deployed. This dependency is critical because the secret does not exist until GitLab is deployed.

```
data "kubernetes_secret" "gitlab_pwd" {
  metadata {
    name      = "gitlab-gitlab-initial-root-password"
    namespace = kubernetes_namespace.gitlab.
                metadata[0].name
  }
  depends_on = [helm_release.gitlab]
}
```

Terraform Apply

Run our usual terraform init followed by terraform apply login at https://github.<domain> with login root and password you get via terraform output gitlab_pwd. Enjoy.

Possible Issues

GitLab is big. It needs some time to get properly initialized. I would recommend watching the process with

`kubectl -n gitlab get po -w`

At the beginning, you will see a lot of pods crashing and restarting. No worries, it takes up to 15 minutes to get alive. If after 15 minutes some pods are still unhealthy, drop them with `kubectl delete po` command.

Conclusion

This Terraform configuration provides a solid initial setup for deploying GitLab in a Kubernetes cluster, ensuring that your instance is secured, organized, and fully operational within the specified domain. However, while this setup is robust, it's important to note that the configuration currently uses the default PostgreSQL database within the Kubernetes cluster. For production environments, you should consider enhancing this setup by migrating the PostgreSQL database to AWS RDS. Using RDS will provide better performance, reliability, and scalability, allowing you to manage your database separately from the cluster, which is crucial for high-availability and disaster recovery scenarios. This improvement will help you fully leverage the benefits of a managed database service, ensuring your GitLab instance is ready to handle increased workloads as your project grows.

CHAPTER 10

Practical Example Part 1

Introduction

It's impossible to predict exactly what kind of application you're going to create, but rest assured—we're prepared for anything. You might need to add another layer or folder where you'll set up a database or other types of storage. However, I want to share how I tackled my own challenges with great success. The solution I developed not only solved my problem but also earned high praise from everyone who used it. They were impressed, and I received a lot of positive feedback. Let's dive into the issue I faced and how I overcame it.

A Learning Environment

For my "Practical Kubernetes" course, I needed an environment that was not only easy to set up and tear down but also provided each student with their own isolated IDE, complete with access to a Kubernetes cluster. The setup had to ensure that one student's work wouldn't interfere with another's, and everything had to run seamlessly in a browser. My goal was to have a simple file where I could input a student's email, run terraform apply, and instantly generate a personalized environment for that student. This environment needed to include an editor, a console with kubectl installed, and all the necessary permissions to interact with the Kubernetes cluster.

A New Folder

Let's create 04-webide folder for our WebIDE inventory, so you would get something like this:

00-vpceks
01-k8s
02-karpenter
03-gitlab
04-webide

Please, cd to 04-webide and create variables.tf.

Content of variables.tf

```
variable "users" {
  description = "List of student emails"
  type = list(any)
}

variable "admins" {
  description = "List of admin user names"
  type = list(any)
}

variable "cluster_name" {
  default = "<your-cluster-name>"
}
```

```
variable "domain" {
  default = "<your-domain>"
}
variable "subdomain" {
  default = "students"
}
variable "storage_class" {
  default = "ebs"
}
variable "tag" {
  default = "webide"   // or whatever you wish
}
// get it via terraform output while you are in 00-vpceks
variable "ecr_url" {
  default = "<ecr-url>"
  description = "Elastic Container URL for our Docker image"
}
```

Content of main.tf

We start in our usual style:

```
terraform {
  backend "s3" {
    bucket = "<your-state-bucket>"
    key    = "04-webide.tfstate"
    region = "<region>"
  }
}
```

CHAPTER 10   PRACTICAL EXAMPLE PART 1

```
provider "aws" {
  region = "<region>"
}

data "aws_region" "current" {}
data "aws_caller_identity" "current" {}

data "aws_eks_cluster" "cluster" {
  name = var.cluster_name
}

data "aws_eks_cluster_auth" "cluster" {
  name = var.cluster_name
}

provider "kubernetes" {
  host                   = data.aws_eks_cluster.cluster.endpoint
  cluster_ca_certificate = base64decode(data.aws_eks_
                           cluster.cluster.certificate_
                           authority.0.data)
  token                  = data.aws_eks_cluster_auth.
                           cluster.token
}
```

Now, let's add a touch of magic to our setup. The list of students is provided in a format like this: ["foo.bar@example.com", "spam.egg@example.com"]. However, the system username must not contain any special symbols. Since all my students were from the same domain, I needed to transform their usernames into a format like foo-bar and spam-egg. To achieve this, I used two Terraform functions: replace() to swap out the special characters, and substr() to extract the relevant part of the email, excluding the domain. These functions allowed me to automatically generate clean, unique usernames for each student, ensuring seamless integration with the system.

```
locals {
  user_names   = [for user in var.users : split("@",
                 replace(user, ".", "-"))[0]]
}
```

Setting Up User Access

There are two key areas where we need to set up user access: AWS itself and the Kubernetes cluster. For AWS, the access requirements are minimal—just enough to get kubectl working. On the Kubernetes side, the access control is a bit more nuanced, but still straightforward. Each user should have full control within their own namespace, with read-only access to the default namespace. Admins, on the other hand, need full access to all Kubernetes resources.

Given that admin users are already in AWS, our primary task for them is to configure their Kubernetes access. To begin, let's create a users.tf file and start with the AWS user creation. This will ensure that each user has the necessary credentials to interact with both AWS and Kubernetes, aligned with their specific permissions.

AWS User Creation

```
resource "aws_iam_user" "users" {
  for_each      = toset(local.user_names)
  name          = each.value
  path          = "/"
  force_destroy = true
}
```

CHAPTER 10   PRACTICAL EXAMPLE PART 1

AWS Access Key

While there might be more secure methods to ensure that a user's WebIDE inherits their permissions, setting AWS keys via environment variables works just fine for my use case. It's straightforward and effective for quickly getting students up and running. Therefore, our next step is to create these keys. This will provide each user with the necessary AWS credentials, ensuring they can interact with AWS resources directly from their WebIDE with the appropriate permissions.

```
resource "aws_iam_access_key" "users" {
  for_each = toset(local.user_names)
  user     = aws_iam_user.users[each.key].id
}
```

Every User Gets an IAM Policy Attached

To ensure that each user has the appropriate level of access, we'll attach a specific IAM policy to each user's account. This policy will define the minimal permissions required to interact with AWS services, specifically tailored to allow the use of kubectl and other essential tools within their WebIDE. By attaching these IAM policies, we maintain a controlled environment where each user's actions are securely confined to their designated permissions, ensuring both functionality and security in the AWS ecosystem.

```
resource "aws_iam_policy" "user_access" {
  name        = "${var.tag}_users_access"
  path        = "/"
  description = "perms for users"
  policy      = <<EOF
```

```
{
    "Version": "2012-10-17",
    "Statement": [
        {
            "Effect": "Allow",
            "Action": [
                "eks:DescribeCluster",
                "eks:ListClusters"
            ],
            "Resource": "*"
        }
    ]
}
EOF
}

resource "aws_iam_user_policy_attachment" "user_access" {
  for_each   = toset(local.user_names)
  user       = each.key
  policy_arn = aws_iam_policy.user_access.arn
  depends_on = [aws_iam_user.users]
}
```

RBAC: Role-Based Access Control for Students

In this section, we're implementing Role-Based Access Control (RBAC) to manage the permissions of our students within the Kubernetes cluster. The goal is to provide each student with specific access rights: read-only access to the default namespace and full control over their own namespaces.

CHAPTER 10   PRACTICAL EXAMPLE PART 1

1. **Cluster Role Binding for Node Viewing**:
   - **Purpose**: We start by creating a kubernetes_cluster_role_binding resource that binds each student to a ClusterRole called node_viewer. This role grants read-only permissions to certain cluster-wide resources like nodes and storage classes.
   - **Explanation**: The node_viewer_binding is created for each student in the list, allowing them to get, list, and watch nodes, storage classes, namespaces, and cluster issuers. This ensures that students have visibility into the cluster's state without the ability to make changes.

2. **Creating the Node Viewer Cluster Role**:
   - **Purpose**: The kubernetes_cluster_role resource defines the node_viewer role, which encompasses the permissions necessary for viewing cluster-wide resources.
   - **Explanation**: This role is crucial for giving students the ability to observe but not modify important resources. It's a key part of ensuring that students can learn and explore without impacting the cluster's integrity.

3. **Namespace Creation for Each User**:
   - **Purpose**: We create a dedicated namespace for each student using the kubernetes_namespace resource.
   - **Explanation**: This ensures that each student has a separate environment to work in, isolating their activities from others and preventing any interference.

4. **Access to the Default Namespace**:
    - **Purpose**: The kubernetes_role_v1 and kubernetes_role_binding_v1 resources provide students with read-only access to the default namespace.
    - **Explanation**: By binding the students to the eks-user-role, they can view resources in the default namespace, such as services or deployments, without being able to modify them. This read-only access is useful for understanding how the cluster operates at a broader level.

5. **Full Access to Own Namespace**:
    - **Purpose**: Finally, we give each student full control over their own namespace through the kubernetes_role_v1 and kubernetes_role_binding_v1 resources.
    - **Explanation**: Each student gets a role specific to their namespace that allows them to perform any action (get, list, watch, create, delete, etc.) on any resource. This unrestricted access within their own namespace encourages exploration and hands-on learning while ensuring they can't affect other students' work.

Put the following content into users.tf:

```
resource "kubernetes_cluster_role_binding" "node_viewer_binding" {
  for_each = toset(local.user_names)

  metadata {
    name = "node-viewer-binding-${each.key}"
  }
```

```
    subject {
      kind      = "User"
      name      = each.key
      api_group = "rbac.authorization.k8s.io"
    }
    role_ref {
      kind      = "ClusterRole"
      name      = kubernetes_cluster_role.node_viewer.
                  metadata[0].name
      api_group = "rbac.authorization.k8s.io"
    }
}
resource "kubernetes_cluster_role" "node_viewer" {
  metadata {
    name = "node-viewer"
  }

  rule {
    api_groups = ["", "storage.k8s.io", "cert-manager.io"]
    resources  = ["nodes", "storageclasses", "namespaces",
                  "clusterissuers"]
    verbs      = ["get", "list", "watch"]
  }
}
resource "kubernetes_namespace" "user_namespace" {
  for_each = toset(local.user_names)
  metadata {
    name = each.key
  }
}
```

```
resource "kubernetes_role_v1" "access_default_ns" {
  metadata {
    name      = "eks-user-role" // not to be confused with
                IAM role
    namespace = "default"
  }
  rule {
    api_groups = ["*"]
    resources  = ["*"]
    verbs      = ["get", "list", "watch"]
  }
}
resource "kubernetes_role_binding_v1" "access_default_ns" {
  for_each = toset(local.user_names)
  metadata {
    name      = "eks-users-role-binding-${each.key}"
    namespace = "default"
  }
  role_ref {
    api_group = "rbac.authorization.k8s.io"
    kind      = "Role"
    name      = kubernetes_role_v1.access_default_
                ns.metadata[0].name
  }
  subject {
    kind      = "User"
    name      = each.key
    api_group = ""
  }
}
```

CHAPTER 10   PRACTICAL EXAMPLE PART 1

```
resource "kubernetes_role_v1" "access_own_ns" {
  for_each = toset(local.user_names)
  metadata {
    name      = each.key
    namespace = resource.kubernetes_namespace.user_
                namespace[each.key].metadata[0].name
  }

  rule {
    api_groups = ["*"]
    resources  = ["*"]
    verbs      = ["*"]
  }
}

resource "kubernetes_role_binding_v1" "access_own_ns" {
  for_each = toset(local.user_names)

  metadata {
    name      = "user-role-binding"
    namespace = resource.kubernetes_namespace.user_
                namespace[each.key].metadata[0].name
  }

  role_ref {
    api_group = "rbac.authorization.k8s.io"
    kind      = "Role"
    name      = each.key
  }
```

```
    subject {
      kind      = "User"
      name      = each.key
      api_group = ""
    }
  }
}
```

Access to Kubernetes

Alright, let's take a moment to refresh your knowledge about configuring access to your EKS Kubernetes cluster. The magic happens through the aws-auth ConfigMap, which is the bridge between AWS IAM and Kubernetes RBAC (Role-Based Access Control). In this chapter, we're going to dive into how we adjust the mapUsers key of the aws-auth ConfigMap to grant the necessary permissions to our users.

To kick things off, we'll create a template file that will define how these user mappings are structured. Go ahead and create a map-users.tpl file with the following content:

```
%{ for admin in admins ~}
- userarn: "arn:aws:iam::${account_id}:user/${admin}"
  username: ${admin}
  groups:
    - system:masters
%{ endfor ~}
%{ for user in users ~}
- userarn: "arn:aws:iam::${account_id}:user/${user}"
  username: ${user}
  groups: []
%{ endfor ~}
```

CHAPTER 10 PRACTICAL EXAMPLE PART 1

As you can see in the template, we're adding admin users to the system:masters group. This group grants full administrative access to the Kubernetes cluster, ensuring that admins can manage all resources without any restrictions. On the other hand, students aren't assigned to any group just yet. We'll handle their permissions separately, giving them just the right amount of access within their own namespaces later on. This setup keeps things balanced—admins get the keys to the kingdom, while students stay safely within their own spaces, only able to do what they need.

Now, to bring this all together, I'd recommend creating an aws_auth.tf file with the following:

```
data "kubernetes_config_map_v1" "aws_auth" {
    metadata {
        name      = "aws-auth"
        namespace = "kube-system"
    }
}

resource "kubernetes_config_map_v1_data" "aws_auth" {
  metadata {
    name      = "aws-auth"
    namespace = "kube-system"
  }

  data = {
    mapRoles = data.kubernetes_config_map_v1.aws_auth.data.mapRoles
    mapUsers = templatefile("${path.module}/map-users.tpl",
      {
        "users"  = local.user_names,
        "admins" = var.admins,
```

```
      "account_id" = data.aws_caller_identity.current.
                     account_id
    })
  }
  lifecycle {
    prevent_destroy = true
  }
  force = true
}
```

Now, let's talk about aws-auth a bit more—it's one of those things that can feel like a riddle wrapped in a mystery, but trust me, it's worth understanding. When you create an EKS cluster, every Node Group gets its own dedicated role, which then finds a cozy spot in the mapRoles section of the aws-auth ConfigMap. In our example, we kicked things off with one Node Group, so after running terraform apply in the 00-vpceks directory, we ended up with one record in aws-auth.

But here's where things get interesting—when we move on to 02-karpenter and run the apply, we're adding another role into mapRoles. This new role is specifically for the Karpenter NodeGroup. Now, you might be asking, "Why is that necessary?" Well, it turns out Karpenter manages nodes with a level of efficiency that EKS just doesn't match on its own. However, if that role isn't included in mapRoles, the EC2 instance created by Karpenter won't be able to join the cluster. Think of it like trying to enter a secure building without the proper credentials—no role in mapRoles, no access.

And now you might be asking, "How's a role connected with an instance anyway?" Great question! It's all linked through the Instance IAM Profile—that's the backstage pass that makes it all happen.

So, back to our code—what we're doing here is preserving the existing mapRoles and updating mapUsers with our list of admins and students. It might sound complex now, but don't worry if it's not crystal clear yet. These things take time to sink in. I spent more than a few late nights puzzling over this before it finally clicked. You'll get there—just keep at it.

We also put prevent_destroy to prevent clearing up aws-auth on terraform destroy. Use `terraform state rm kubernetes_config_map_v1_data.aws_auth` if you ever wish to run terraform destroy in that folder.

And here's a little nugget of wisdom that's definitely worth mentioning: the account from which your EKS cluster was created—yep, that very account—isn't listed in aws-auth, yet it always has full access to the cluster. It's like having a master key tucked away in your pocket. This is a fantastic safety net because no matter what happens, even if aws-auth gets totally screwed up, you can still access and repair the cluster. It's a comforting thought, knowing that you've got this built-in backup, keeping things from going off the rails completely. So, even when the skies are cloudy, you've always got a way to bring the sunshine back to your cluster.

CHAPTER 11

Practical Example Part 2

DNS Records

We want our students to access the system via a unique URL that looks like this: <user-name>.students.<domain>. Since we have a hosting zone created in the 00-vpceks folder, here, we just need to fetch data about it. So, create dns.tf and put the following:

```
data "aws_route53_zone" "parent" {
  name = var.domain
}
resource "aws_route53_zone" "this" {
  name = "${var.subdomain}.${var.domain}"
}
```

This will create a hosting zone for students.<domain>. Now we need to set up an NS record in the parent zone so our setup actually works.

```
resource "aws_route53_record" "parent" {
  zone_id = data.aws_route53_zone.parent.zone_id
```

CHAPTER 11 PRACTICAL EXAMPLE PART 2

```
  name    = aws_route53_zone.this.name
  type    = "NS"
  ttl     = "300"
  records = aws_route53_zone.this.name_servers
}
```

Once routing is set up, we need an SSL certificate that will work for *.students.<domain>. Here, we go

```
resource "aws_acm_certificate" "this" {
  domain_name = aws_route53_zone.this.name
  //additional domains wild
  subject_alternative_names = ["*.${aws_route53_zone.
  this.name}"]

  validation_method = "DNS"
}
resource "aws_route53_record" "validation" {
  for_each = {
    for dvo in aws_acm_certificate.this.domain_validation_
    options : dvo.domain_name => {
      name   = dvo.resource_record_name
      record = dvo.resource_record_value
      type   = dvo.resource_record_type
    }
  }

  allow_overwrite = true
  name            = each.value.name
  records         = [each.value.record]
  ttl             = 60
  type            = each.value.type
  zone_id         = aws_route53_zone.this.zone_id
}
```

```
resource "aws_acm_certificate_validation" "this" {
  certificate_arn         = aws_acm_certificate.this.arn
  validation_record_fqdns = [for record in aws_route53_record.
validation : record.fqdn]
}
```

Docker Image

Please create Dockerfile with the following content:

```
FROM codercom/code-server:latest
```

```
RUN  sudo apt update && sudo apt install -y python3-pip wget
unzip && sudo pip3 install awscli --break-system-packages
```

```
RUN curl -LO "https://dl.k8s.io/release/$(curl -L -s https://
dl.k8s.io/release/stable.txt)/bin/linux/amd64/kubectl" && \
   sudo mv kubectl /usr/local/bin && sudo chmod +x /usr/local/
bin/kubectl
```

```
CMD ["dumb-init", "/usr/bin/code-server", "--bind-addr",
"0.0.0.0:8080"]
```

Here, we are using a very well-done code server from Coder which let us have Visual Studio Code inside browser and installing kubectl.

Build and Push Docker Image

Get the ECR url from `terraform` output run in 00-vpceks and run the docker build command:

```
docker build -t <ecr-url>:latest .
```

If you are building on a Mac M1, add `--platform linux/amd64` to your build command. Do not forget about "." at the end.

We need to let our local docker service to be authorized before pushing to ECR. Please run

```
aws ecr get-login-password --region <region> | docker login --username AWS --password-stdin <ecr-url>
```

And push

```
docker push <ecr-url>:latest`
```

Namespaces for WebIDE

Let's create a `webide.tf` file (or whatever name you prefer) and add the following:

```
resource "kubernetes_namespace" "webide" {
  metadata {
    name = "webide"
  }
}
```

Persistent Storage for Student Work

Life in the cloud is dynamic—things are always happening simultaneously. Given that we're using SPOT instances for our WebIDE, restarts are to be expected. Karpenter does a great job of handling these smoothly, but we still need to ensure that a student's work is preserved through these disruptions. To do this, we'll create a 1 GB home directory that will be mounted to the student environment. This directory will be backed by a PersistentVolumeClaim (PVC), ensuring that each student's data is stored reliably and remains accessible even if their instance is restarted.

```
resource "kubernetes_persistent_volume_claim" "home" {
  for_each = toset(local.user_names)
  metadata {
    name      = "home-${each.key}"
    namespace = kubernetes_namespace.webide.metadata.0.name
    annotations = {
      "volume.kubernetes.io/storage-provisioner" =
      "ebs.csi.aws.com"
    }
  }
  spec {
    access_modes = ["ReadWriteOnce"]
    resources {
      requests = {
        storage = "1Gi"
      }
    }
    storage_class_name = var.storage_class
  }
}
```

Service for WebIDE

The rule is simple: if our app is exposed via the network, we need a service for that. Let me explain—while pods in Kubernetes receive IP addresses, those IPs are not persistent. This means that as pods are created and destroyed, their IPs change. On the other hand, a service in Kubernetes gets a static IP address from Kubernetes' own IP range. This static IP, along with a DNS name, can be used to consistently access the pods behind the service, no matter how many times they are restarted or replaced. This ensures reliable network access to your application, whether it's for internal communication within the cluster or for external access.

```
resource "kubernetes_service" "webide" {
  for_each = toset(local.user_names)
  metadata {
    name      = "webide-${each.key}"
    namespace = kubernetes_namespace.webide.metadata.0.name
  }
  spec {
    selector = {
      app = "webide-${each.key}"
    }

    port {
      port        = 8080
      target_port = 8080
    }
  }
}
```

In this step, we're creating a service that load balances traffic to any pod with the app:webide- label. While this service is designed to handle multiple pods, in our case, it will only be managing one pod per student. This effectively makes the service act as a proxy, ensuring that traffic is routed to the correct WebIDE instance for each student.

Introducing Ingress

In a Kubernetes environment, managing how external traffic reaches your services is crucial. This is where Ingress comes into play. An Ingress is a Kubernetes resource that allows you to define rules for accessing your services from outside the cluster. Instead of exposing each service individually with its own Load Balancer (LB), which can quickly become expensive and cumbersome, Ingress lets you route all traffic through a single Load Balancer, dramatically simplifying your infrastructure.

One of the key reasons to use Ingress is the need for SSL (Secure Sockets Layer) to secure your web traffic. With Ingress, you can terminate SSL at the Load Balancer level, ensuring that all communications between your users and the cluster are encrypted and secure. This not only enhances security but also centralizes the management of SSL certificates, making it easier to maintain and renew them.

Another significant advantage of using Ingress is the ability to integrate with ExternalDNS, a Kubernetes project that automatically manages DNS records for your services. With ExternalDNS, you can automate the process of updating DNS records whenever your services change, ensuring that your domain names always point to the correct IP addresses without manual intervention.

Creating Ingress for Every Student

```
resource "kubernetes_ingress_v1" "webide" {
  for_each = toset(local.user_names)
  metadata {
    name      = "webide-${each.key}"
    namespace = kubernetes_namespace.webide.metadata.0.name
    annotations = {
      "kubernetes.io/ingress.class"                  = "alb"
      "alb.ingress.kubernetes.io/scheme"             =
      "internet-facing"
      "alb.ingress.kubernetes.io/target-type"        = "ip"
      "alb.ingress.kubernetes.io/certificate-arn"    =
      aws_acm_certificate.this.arn
      "alb.ingress.kubernetes.io/listen-ports"       =
      "[{\"HTTP\": 80}, {\"HTTPS\": 443}]"
```

CHAPTER 11   PRACTICAL EXAMPLE PART 2

```
        "alb.ingress.kubernetes.io/group.name"     =
        "${var.subdomain}-${var.domain}"
        "alb.ingress.kubernetes.io/ssl-redirect"    = "443"
        "alb.ingress.kubernetes.io/healthcheck-path" = "/login"
      }
    }
    spec {
      ingress_class_name = "alb"
      rule {
        host = "${each.key}.${var.subdomain}.${var.domain}"
        http {
          path {
            backend {
              service {
                name = "webide-${each.key}"
                port {
                  number = 8080
                }
              }
            }
            path      = "/*"
            path_type = "ImplementationSpecific"
          }
        }
      }
    }
  }
}
```

A few notes: By using alb.ingress.kubernetes.io/group.name, we can utilize a single Load Balancer for all resources instead of creating a separate Load Balancer for each student. To ensure this works correctly, we had to apply the appropriate tags to our subnets. You can refer to the `vpc.tf` file in the `00-vpceks` folder for details. Behind the scene ExternalDns that we set in 01-k8s puts the correct CNAME record for the student's personal hostname.

Init Script

I want my students to spend zero time setting up their environments. This means that kubectl should be ready to go and configured to point to the student's namespace by default, at a minimum. We might want to add more adjustments in the future as needed. To achieve this, we'll create an initialization script that will be stored in a ConfigMap. This ConfigMap will be mounted to the student's environment and executed during the initialization process.

```
resource "kubernetes_config_map" "this" {
  for_each = toset(local.user_names)
  metadata {
    name      = "setcontext-${each.key}"
    namespace = kubernetes_namespace.webide.metadata.0.name
  }
  data = {
    "setcontext.sh" = "aws eks update-kubeconfig --name ${var.
    cluster_name} && kubectl config set-context --current
    --namespace=${kubernetes_namespace.user_namespace[each.
    key].metadata[0].name}"
  }
}
```

You might consider to use templatefile function, for example, `"setcontext.sh" = templatefile(...)`.

CHAPTER 11   PRACTICAL EXAMPLE PART 2

Random Password

Every student is provided with a link to their environment along with a randomly generated password. While this setup might not be secure enough for an enterprise solution, it works perfectly well for my one-week training course. The password ensures that each student's environment is protected, while still being easy to manage and distribute.

To generate these passwords, we use a random_id resource in Terraform. This resource creates a unique token for each student, ensuring that their environment is secure without the need for complex password management.

```
resource "random_id" "token" {
  for_each    = toset(local.user_names)
  byte_length = 16
}
```

CHAPTER 12

Practical Example Part 3

Deploying the WebIDE with StatefulSet

Next, we're deploying the WebIDE environment for each student using a kubernetes_stateful_set_v1 resource. StatefulSets are ideal for managing stateful applications like our WebIDE, ensuring that each instance has a stable network identity and persistent storage.

- **Metadata**: Each StatefulSet is uniquely named based on the student's username and is associated with the corresponding namespace. Labels are used to match the Pods to their corresponding services.
- **Spec**:
  - **Replicas**: We're running a single replica of the WebIDE for each student.
  - **Service Name**: Each StatefulSet is associated with a service, ensuring stable networking for the WebIDE.

- **Volume Configuration:**
    - The home volume is mounted to the /home/coder directory, backed by the previously defined PVC.
    - An entrypoint volume is also configured to mount a ConfigMap containing an initialization script, which is executed during the container's startup.
- **Init Container:**
    - The init_container is used to set up the environment before the main container starts. It ensures that the home directory has the correct permissions and executes the initialization script (setcontext.sh).
    - **Environment Variables:** AWS credentials and region information are passed as environment variables to ensure that the container can interact with AWS services securely.
    - **Volume Mounts:** The home directory and the initialization script are mounted to the appropriate paths.
- **Main Container:**
    - The main container runs the WebIDE environment, pulling the specified Docker image and configuring resource limits to ensure stable operation.
    - **Environment Variables:** In addition to the AWS credentials, a randomly generated password (created earlier) is passed to secure the WebIDE.
    - **Volume Mounts:** The home directory is mounted to the same path as in the init container, ensuring that student data persists across sessions.

```
locals {
    webide_image = "${var.ecr_url}:latest"
}
resource "kubernetes_persistent_volume_claim" "home" {
  for_each = toset(local.user_names)
  metadata {
    name      = "home-${each.key}"
    namespace = kubernetes_namespace.webide.metadata.0.name
    annotations = {
       "volume.kubernetes.io/storage-provisioner" = "ebs.csi.aws.com"
    }
  }
  spec {
    access_modes = ["ReadWriteOnce"]
    resources {
      requests = {
        storage = "2Gi"
      }
    }
    storage_class_name = var.storage_class
  }
}
resource "kubernetes_stateful_set_v1" "webide" {
  for_each = toset(local.user_names)
  metadata {
    name      = "webide-${each.key}"
    namespace = kubernetes_namespace.webide.metadata.0.name
    labels = {
      app = "webide-${each.key}"
    }
  }
```

```
spec {
  replicas = 1
  selector {
    match_labels = {
      app = "webide-${each.key}"
    }
  }
  service_name = "webide-${each.key}"
  template {
    metadata {
      labels = {
        app = "webide-${each.key}"
      }
    }
    spec {
      volume {
        name = "home"
        persistent_volume_claim {
          claim_name = kubernetes_persistent_volume_claim.
                    home[each.key].metadata[0].name
        }
      }
      volume {
        name = "entrypoint"
        config_map {
          name = kubernetes_config_map.this[each.key].
              metadata[0].name
        }
      }
      init_container {
        image    = local.webide_image
```

```
    name    = "chown"
    command = ["/bin/sh", "-c", "sudo chown 1000:1000 -R
              /home/coder && sh /opt/setcontext.sh"]
    env {
      name  = "AWS_DEFAULT_REGION"
      value = data.aws_region.current.name
    }
    env {
      name  = "AWS_ACCESS_KEY_ID"
      value = aws_iam_access_key.users[each.key].id
    }
    env {
      name  = "AWS_SECRET_ACCESS_KEY"
      value = aws_iam_access_key.users[each.key].secret
    }
    volume_mount {
      name       = "home"
      mount_path = "/home/coder"
    }
    volume_mount {
      name       = "entrypoint"
      mount_path = "/opt/setcontext.sh"
      sub_path   = "setcontext.sh"
    }
  }

  container {
    image             = local.webide_image
    name              = "main"
    image_pull_policy = "Always"

    resources {
      requests = {
```

CHAPTER 12 PRACTICAL EXAMPLE PART 3

```
          memory = "500Mi"
        }
        limits = {
          memory = "1Gi"
        }
      }
      env {
        name  = "AWS_DEFAULT_REGION"
        value = data.aws_region.current.name
      }
      env {
        name  = "AWS_ACCESS_KEY_ID"
        value = aws_iam_access_key.users[each.key].id
      }
      env {
        name  = "AWS_SECRET_ACCESS_KEY"
        value = aws_iam_access_key.users[each.key].secret
      }
      env {
        name  = "PASSWORD"
        value = random_id.token[each.key].hex
      }
      volume_mount {
        name       = "home"
        mount_path = "/home/coder"
      }
    }
  }
 }
 }
}
```

CHAPTER 12 PRACTICAL EXAMPLE PART 3

Access Details Distribution

Once we have the environments created, how do we distribute the access details to each student in an easily accessible format? The most convenient solution for us turned out to be generating files directly from Terraform that contain the access details for each student. This approach ensures that each student receives their credentials in a straightforward, easy-to-distribute format.

To achieve this, we start by creating a template file student_secret.tpl with the following content:

```
# WebIDE
https://${user}.${domain}?folder=/home/coder
Password: ${password}
```

This template will be used to generate personalized access details for each student.

Next, update webide.tf with the following:

```
resource "local_file" "secrets" {
  for_each = toset(local.user_names)
  filename = "${path.module}/secrets/${each.key}.txt"
  content  = templatefile(
    "${path.module}/student_secret.tpl",
    {
      user     = each.key
      domain   = "${var.subdomain}.${var.domain}"
      password = random_id.token[each.key].hex
    }
  )
}
```

During the terraform apply, this configuration will generate one file per user in the `./secrets` folder, containing their unique access details. Each file includes the WebIDE URL, specific to the user, along with their login password. This method not only simplifies distribution but also ensures that every student has quick and secure access to their environment.

Providing List of Students

The last piece of the puzzle is finding a convenient way to provide the list of users and admins. To do this, we create `terraform.tfvars` file with the following content:

```
admins = ["<trainer-username>"]
users = [
    "foo.bar@example.com",
    "spam.egg@example.com",
    ...
]
```

The file must be named exactly `terraform.tfvars` because this is a standard way in Terraform to provide input variable values. By using this file, you can easily manage and update the list of users and admins for your training sessions. This approach simplifies the process of setting up your environment, making it easy to scale up or down as needed, and ensures that all variables are consistently and correctly applied during the Terraform run.

Further Improvements

We took things a step further by generating SSH keys for every student using the tls_private_key resource. These keys were then used to set up student access to our GitLab instance via the GitLab provider. This enhancement not only secured access to the GitLab repositories but also provided a seamless integration experience for the students. However, I'm leaving this exercise for you. By implementing this, you'll ensure that students can securely push and pull code from GitLab, enhancing their learning experience even further.

Time to Apply

It's time for you to take the knowledge you've gained from this book and put it into action. Run that `terraform apply`, and with it, you're bringing to life the powerful infrastructure you've meticulously crafted. Congratulations—you've reached a significant milestone.

Conclusion

The solutions in this book weren't invented in a single burst of inspiration; they revealed themselves through countless iterations, trials, and improvements. What you hold in your hands is the 100th version of an EKS cluster setup and the 21st approach to deploying GitLab—each version a step closer to perfection. These are the distilled results of hard-earned experience, refined to give you the most effective tools and strategies. And it is still not perfect. I am looking forward for version 200.

As you move forward, remember that you're equipped with a great set of skills. The infrastructure you've built isn't just a collection of code and configurations; it's a testament to your dedication, adaptability, and commitment to excellence. Enjoy the journey ahead, confident in the knowledge that you've mastered the art of creating robust, scalable, and efficient systems. Your next challenge awaits, and with the skills you've honed, you're ready to conquer it.

Index

A

Amazon Certificate Manager (ACM), 101, 102
Application layer, 2
AWS Identity and Access Management (IAM)
    karpenter policy, 21–24, 26
    policy, 19–21
    reducing computer bills, 26–28
    role, 20, 21
AWS's Application Load Balancer (ALB), 69

B

Bitbucket, 91

C

cAdvisor, 40
CI/CD pipelines
    GitLab actions, 91
cidrsubnet function, 15
CoreDNS, 54
Custom Resource Definition (CRD), 77

D

Directory structure, 2, 3
Docker image
    build/push, 125
    create file, 125
    creating ingress, 129, 130
    ingress, 128
    init script, 131
    persistent storage, 126, 127
    random password, 132
    WebIDE, namespaces, 126
    WebIDE service, 127, 128
Domain Name System (DNS), 15, 17, 123, 124

E, F

EKS Add-ons
    building and scaling applications, 54
    components, 53
    CoreDNS, 54
    kube-proxy, 54
    VPC-CNI, 54
Elastic Container Registry (ECR), 1, 17, 18
Elastic Kubernetes Service (EKS), 1, 44

G

GitLab, 140, 141
   components, 92
   configuration, 97–100
   DNS records, 103
   girlab.tf
      Helm, 102, 103
      Kubernetes namespace, 101
      SSL certificate, 101
   Helm, install, 92, 93
   main.tf, 93, 94
   output.tf, 95
   root password, 104
   terraform, 104
   *.tpl, 95
   values file, 97
   values.tpl, 96
   variables.tf, 94
GitLab cluster, 2
GitOps principles, 57

H

Helm
   advantages, 56
   automation, 57
   aws-auth ConfigMap, 79, 80, 89, 90
   backend configuration, 80, 81
   charts, 55
   CLI, 55–57
   cluster information/kubernetes provider, 81, 82
   consistency, 57
   customization, 56
   EC2NodeClass, 88
   GitOps, 57
   hardparts
      aws-auth ConfigMap, 78
      CRD, 77, 78
      karpenter, 78
      kubectl commands, 76
      kubectl configuration, 76
      kubectl api-resources, 77
      Kubernetes cluster, 75
   IAM rol, policies, 84, 85
   Karpenter, creating resources, 85–87
   Karpenter nodes, IAM roles, 82, 83
   Red Hat Linux, 55
   releases, 56
   repositories, 55
   sharing and reuse, 56
   simplifies deployments, 56
   version control, 56, 57

I, J

Ingress Controller, 68
Ingress object, 68
init_container, 134

K, L, M

kubeadm, 43
kubectl command, 80

Kubelet, 39
Kube-proxy, 40, 54
Kubernetes
    ALI ingress controller, 69
    AWS, setting up cluster, 43
    cluster, 43
    control hub, 31–33
    controller manager, 35–38
    database, etcd, 29–31
    default storage class, 63, 64
    EBS, custom storage class, 60
    EKS, 44
    external DNS, 70–73
    Ingress controller, 68
    Ingress object, 68
    Karpenter application, 64, 66, 67
    nameservers, 48
    servers, 33–35
    storage class, 59
    terraform code, storage, 61–63
    terraform EKS module, 44–48
    worker nodes, 39–41
Kubernetes API Server, 32
Kubernetes operations (kops), 44
Kubernetes resources, 1
    EKS Add-ons, 54, 55
    Helm, 55–57
    main.tf
        AWS provider, 52
        data sources, 52
        Helm provider, 53
        Kubernetes provider, 52
    state and providers, 50, 51
    terraform block
        configuration, 52
    variables, 49, 50

N, O

node_viewer, 114

P, Q

PersistentVolumeClaim (PVC), 40, 59, 126
Persistent Volumes (PVs), 40, 63
Pod, 40
PostgreSQL database, 105
Practical Kubernetes course
    access, 119–122
    access user control, 111
    AWS access key, 112
    IAM policy, 112, 113
    isolated IDE, 107
    main.tf, 109, 110
    RBAC, 113–116, 118, 119
    setting up user access, 111
    variables.tf, 108, 109
    WebIDE inventory, 108

R

Role-Based Access Control (RBAC), 113, 119

INDEX

S

S3 buckets, 1
Secure Sockets Layer (SSL), 128

T, U

Terraform, 53, 101
  access detail distribution, 138, 139
  building VPC, 13–15
  CLI, create state bucket, 11
  create terraform.tfvars file, 140
  declaring resource, 4
  DNS, 15, 16
  ECS, 17, 18
  fetching data, 5
  infra git repo, 11
  infrastructure, 3
  input variables, 6
  local variables, 6
  Main.tf, 12, 13
  output, 7
  providers, 5
  revolution, 4
  setting up CLI, 10
  state management, 4, 5
  variables.tf, 12
  web console, minimal work required, 10
  working directory, 7
Terraform template files (*.tpl), 95

V

Virtual private cloud (VPC), 1, 13
VPC-CNI, 54

W, X, Y, Z

Web Application Firewall (WAF), 69
WebIDE environment, 133, 134, 136–138
WebIDE inventory, 108

GPSR Compliance

The European Union's (EU) General Product Safety Regulation (GPSR) is a set of rules that requires consumer products to be safe and our obligations to ensure this.

If you have any concerns about our products, you can contact us on

ProductSafety@springernature.com

In case Publisher is established outside the EU, the EU authorized representative is:

Springer Nature Customer Service Center GmbH
Europaplatz 3
69115 Heidelberg, Germany

www.ingramcontent.com/pod-product-compliance
Lightning Source LLC
LaVergne TN
LVHW010342260326
834688LV00036B/841